30 ANIMALS
THAT MADE US SMARTER

Dedicated to the curious minds of all creeds and cultural backgrounds. To those who seek out the uncharted maps of discovery to expand the horizons of the mind. Be guided not by the limitations of what the known world bestows upon you, but rather by the power of your imagination that lies deep inside.

3 1526 05722065 6

30 ANIMALS THAT MADE US SMARTER

Stories of the natural world that inspired human ingenuity

PATRICK ARYEE

With Michael Bright

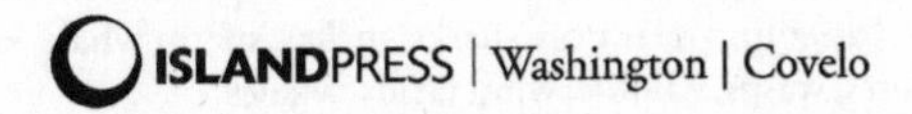

Licensed by BBC News World Service © BBC 2021

First published in the United States by Island Press in 2022

Copyright © 2021 Patrick Aryee

Illustrations © 2021 Lizzie Harper

Patrick Aryee has asserted his right to be identified as the author of this Work.

First published by Woodlands Books Limited, an imprint of Ebury Publishing, a division of The Random House Group Limited.

Based on the BBC World Service podcast 30 Animals That Made Us Smarter, an original idea by James Cook and Jon Manel, produced by Sarah Blunt, Joanna Jolly, James Cook, and Dimitri Houtart.

Licensed by the BBC.

"BBC" and "World Service" are trademarks of the British Broadcasting Corporation and are used under license.

BBC logo © BBC 2021. BBC World Service © BBC 2021.

All rights reserved under International and Pan-American Copyright Conventions. No part of this book may be reproduced in any form or by any means without permission in writing from the publisher: Island Press, 2000 M Street, NW, Suite 480-B, Washington, DC 20036-3319.

Library of Congress Control Number: 2021953417

All Island Press books are printed on environmentally responsible materials.

Manufactured in the United States of America
10 9 8 7 6 5 4 3 2 1

Keywords: adaptations, ants, anatomy, animals, bats, biomimicry, bullet train, butterflies, coral reef, elephant, engineering, evolution, fish, hedgehog, history, innovation, invention, kingfisher, lobster, mantis shrimp, medicine, mosquitoes, mussels, octopus, pangolin, polar bear, sharks, snakes, sperm whale, tardigrade, technology, termites, wasps, wildlife, wind farms, whales

Contents

Blueprints

Our planet is a perfectly calibrated operating system. It's also the only place we know of in the entire universe that's capable of supporting life. But don't be fooled: living on a rock that's hurtling through the solar system at 67,000 miles an hour is fraught with danger. For all its beauty, many of Earth's ecosystems are extreme places to live. From the crushing pressures of the oceans to the vast scorching desert sands; the deathly thin air of mountaintops to the freezing chill of the polar regions. These extreme environments present a major challenge to any and every form of life. Looking around the globe today, however, we find a stunning variety of species, resolute in their multitude of solutions in overcoming these challenges to thrive in the most unlikely of environments.

This sheer tenacity of life comes down to one thing: genetic variation, aka evolution. It's Charles Darwin, the English naturalist, and all-round big dog in the biology game, who is largely credited for theorising that all species arise, and consequently develop, through a process called natural selection. Now stick with me here because it's this 'natural selection' of small inherited variations – or genetic mutations – that is responsible for providing traits and characteristics that increase an organism's ability to survive, compete and reproduce. One standout example of this can be seen in the evolutionary history of one of our most loved and iconic wild animals: the polar bear.

Despite their dazzling appearance, polar bears are in fact descended from a more 'grizzly-looking' brown bear. It was during the last ice age that having white fur instead of brown became more of an advantage. The few bears that found themselves with the genetic mutation that resulted in white *fur* were now camouflaged against the white *snow*. This made it harder for their prey – mainly ringed seals – to spot their predators. This camouflage started out as a random mutation, but it turned into the perfect adaptation, a trait that became naturally selected for and passed down the generations over thousands of years to this day.

This is evolution at its finest. But then again, what if one of these variations were to give you or me, or any animal for that matter, a quality that didn't work so well in helping us to survive? Well, in that case, it's pretty straightforward: it's off to the chopping block. The chances are you'd eventually die and cease to exist along with the rest of your species. It may seem harsh, but life on Earth is all about survival of the fittest.

So here's the million-dollar question. What does all of this have to do with making us smarter? Well, since that miraculous moment when life first began, plants, animals, fungi and bacteria have been constantly working out the best ways to survive and, importantly, adapt to the complex ecosystems of the planet. So, given that nature has had all this time – quite literally billions of years of trouble-shooting – do we find ourselves in a prime position to learn a thing or two ourselves? Well, this is exactly what I set out to investigate in my companion podcast series with the BBC World Service: to explore the remarkable things animals do to survive and see if those examples might help us in overcoming human challenges of the modern era. You see, the natural world is full of problem solvers; after all, many species have adopted all manner of solutions over time, through trial and error. You can think of nature as the ultimate research and development lab. This stuff is called biomimicry, by the way. I've been fascinated by it for as long as I can remember, and I know for a fact you will be too.

What is biomimicry? Simply put, it's the process of designing brand new tech, inspired by the animals of planet Earth. If it's your first time hearing about it, trust me, it's a really cool subject, and as you make your way through the stories in the chapters that follow, I've no doubt that your view of the natural world won't be the same again.

This idea of looking to Mother Nature for inspiration was largely popularised by scientist, author and self-proclaimed 'nature nerd' Janine Benyus, in her 1997 book *Biomimicry: Innovation Inspired by Nature*. She describes biomimicry as a 'new science that studies nature's models [...] to solve human problems.' Janine believes it's imperative for us to see nature as a 'mentor,' and interestingly she even had the foresight to identify sustainability as an inbuilt objective of biomimicry. Today, of course, the need for more sustainable processes and products is more relevant to our survival as a species than ever before.

As a forward-thinking academic, Janine is one of many who've championed this approach to engineering solutions to everyday challenges. Almost 30 years earlier, American biophysicist Otto Herbert Schmitt first used the synonymous term 'biomimetics' to identify this fledgling science, with one of his colleagues Jack E. Steele referring to this same area of research as 'bionics.' One of the earliest examples of biomimicry, however, comes from the Renaissance Man himself: Leonardo da Vinci was absolutely obsessed with flight and is well known for having drawn several sketches of 'flying machines,' the wings of which were based on the wings of a group of animals that, when it comes to aerial manoeuvrability, have complete mastery over the skies – bats. Although he didn't see his visions of the future materialise in his own time, these blueprints undoubtedly served as inspiration for the Wright Brothers centuries later, in what would be a landmark achievement for all humankind: the successful creation of the world's first powered aircraft.

This is in essence what *30 Animals That Made Us Smarter* aspires to shed more light on. How we can learn from animals in the wild and the 3.8-billion-year-old blueprints of Mother Nature. Personally, I love this combination of science, technology, biology and innovation. It's exciting, thought-provoking, and in a world increasingly in need of solutions that combat the effects of climate change, it just makes sense. So, what should you expect in the pages that lie ahead? Well, I'd like to think there are some surprising stories to sink your teeth into, but I'll let you be the judge of that. How about the cows that have inspired new eco-friendly sewage systems; or the mantis shrimps that are revolutionising future aircraft safety; and then there's the manta rays tipped as being the solution to microplastic pollution. All of this, plus X-ray space telescopes inspired by the eyes of lobsters, and jumping robots based on the snapping jaws of a very special group of ants.

One story that certainly captured both my imagination and attention was the research of Frank Fish, a professor of marine biology who became entranced by the anatomy of humpback whales. He found himself perplexed after learning that the leading edges of the whales' flippers weren't smooth as he'd expected, but were in fact covered in huge bumps. As a scientist familiar with the laws of aero- and hydrodynamics, he just couldn't seem to work out why their pectoral flippers had evolved to be like this. It would surely make more sense to be smooth and streamlined like the wings of a plane. I can't give away too much, but what I can say is that his curiosity in asking such a seemingly simple question is right now changing the way we harness renewable energy, in a way that's both efficient and intelligent.

Now we're fully acquainted with the subject of this book, I think the only thing that's left is to introduce myself. Hi, I'm Patrick Aryee. I'm a TV presenter and wildlife filmmaker, and I've always been fascinated by how things work. I first started making wildlife documentaries with the BBC Natural History Unit in 2013 and, since then, I've witnessed some of the most breathtaking sights nature has to offer. Alongside innovative tales of biomimicry, it's been a pleasure to relive the delights of these adventurous moments with you in greater detail. From freediving with giants of the sea to milking deadly rattlesnakes for venom, each of these personal encounters serves to breathe a unique sense of life into the pages and transport you on a wild adventure across the globe.

Oh, and if that wasn't enough, keep an eye out for the hand-crafted illustrations of Lizzie Harper. Our collaboration was fuelled by the desire to create imagery that's moreish to the mind and immerses us as readers into the hidden worlds of these biological superstars. Whether you're an animal lover or just someone with an all-round curiosity for the world around us, there's a little something for everyone. So what are you waiting for? Thirty animals that made us smarter and the blueprints to a brighter future await your discovery...

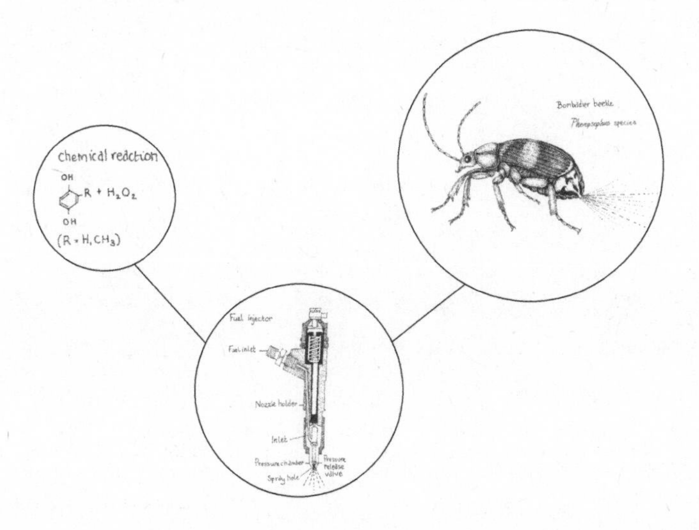

1

The Kingfisher and the Bullet Train

Our first story comes from Japan, and it tells how a bird – the kingfisher – helped to reshape the daily commute of millions. If you were to ask any birdwatcher about how they feel on a day they get to see this brightly coloured bird, don't listen to a word they say; instead, watch for the expressions of joy that flush across their face as they imagine that fleeting moment. Common kingfishers like to make their home in dense cover, close to still or slow moving water. This is the perfect location to more easily find their favourite snack, which, if you haven't already guessed, is all in the name. They're after fish. If you're lucky enough to be in the right place and at the right time, you just might catch a glimpse out the corner of your eye of a sudden flash of bright blue as one darts across the water's surface. If you're *really* patient, you might even see one fishing or perching proud in the sunshine with its catch in its bill.

Someone who's very familiar with kingfishers is Eiji Nakatsu. Eiji, as you might imagine, is a birdwatcher, but he also happens to be an engineer. As the General Manager of the Technical Development Department of Japan Railway West, he was instrumental in the design of the 500 Series Shinkansen, or, as it's more commonly known, the 'Bullet Train.' If you're yet to see them, bullet trains are impressive machines, straight out of the future. They carry millions of passengers on some of the world's busiest railways, and are among the fastest trains on the planet. Back in 1990, though, the trains had a number of issues that Eiji Nakatsu and his research team were challenged to overcome. The first step was to make them even faster. Normally a trip from Shin-Osaka to Hakata station in Fukuoka would take four or five hours, depending how often the service stopped along the way, but the new trains had to drop this journey time so that the same trip would take less than 2 hours and 20 minutes, and, to do that, each train would need to reach speeds of over 185 mph. It would be easy to assume that all they needed to do was jam in a bigger and more powerful engine, but it wasn't power that was the limiting factor.

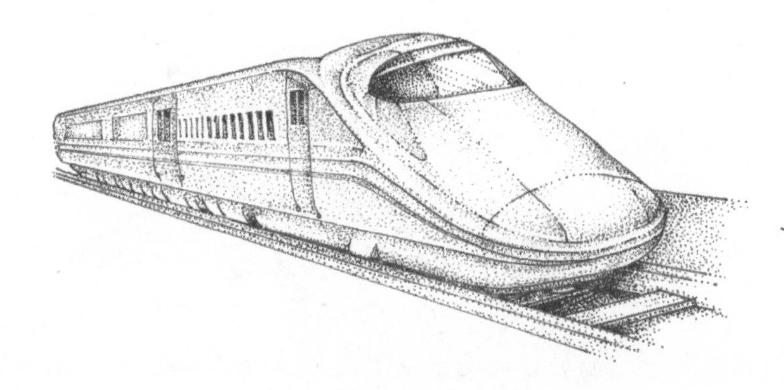

High speed Shinkansen electric train

When the speeding bullet trains went through tunnels, they made a really loud boom each time they came out. This booming sound disturbed both the nearby wildlife and the passengers on board, plus it woke up people who lived in neighbouring houses. Teams of engineers worked around the clock to find a solution. They discovered that it all came down to aerodynamics: as the trains entered the tunnels, air pressure built up in front of them. The best way to visualise this is to think of a bicycle pump: each time the plunger is pushed forwards, air compresses at the front end. The same thing was happening with the trains, only each time they came out of a tunnel, the compressed air was suddenly free to expand... and BOOM!

The noise the trains were creating was so loud, that it repeatedly broke Japanese environmental regulations. To

meet these, the trains had to be quieter than 75 decibels, which is about the same volume as a flushing toilet. When you need a train to travel at over 185 mph, this is seriously quiet. The compressed air was also acting like an invisible brake, slowing down the trains. If the research team could solve these problems, they'd succeed in making the trains both faster and quieter. In short, they needed a train that could slice through the air more efficiently, and that is where our kingfisher comes in.

Our planet is home to about 100 different species of kingfisher. Save for the polar regions and some of the world's driest deserts, you can find them all over the globe, with most species living in Africa, Asia and Australia. They're generally characterised by long, sharp, pointed bills, along with big heads, short legs and stubby tails. Despite their name, not all kingfishers catch fish, but those that do are really efficient at it. Their skill as avian anglers comes down to exceptional eyesight and a beak that's perfectly shaped for plunging into the water. Long, narrow and highly streamlined, the beak steadily increases in diameter from the tip to the head. This adaptation reduces the force of the initial impact, at the moment the kingfisher hits the water's surface at speeds approaching 25 mph. The bird essentially slides into the water, allowing the liquid to flow past the bill, instead of being pushed in front of it. As the bullet trains moved from the open track and into a tunnel, that's exactly what they were failing to do.

As he mulled over this conundrum, Eiji Nakatsu just so happened to attend a lecture by an aviation engineer at the Wild Bird Society of Japan in Osaka. It should've been self-evident, perhaps, but he was genuinely surprised to find out that birds, far from just inspiring manned flight, were still helping aviation specialists to solve some of their biggest design problems. It got him thinking: if the front of his train looked more like a kingfisher beak, could it improve the aerodynamics, allowing air to flow around it more easily and reduce the build-up in front? If it did, it would finally get rid of the boom.

It turned out he was right. Eiji and his team got to work, studying the kingfisher in detail. They realised the cross-section of the upper and lower bill was more complicated than it first appeared. It resembled two triangles, with rounded edges, which together formed a squashed diamond shape. Eiji described it as 'a circular lozenge surrounded by four circles', which, I have to admit, I always find hard to visualise. The crucial point here is that they designed a new nose for the bullet trains that looked more like the beak of a kingfisher. As you can imagine, it looked a little bit odd, especially as it was more than twice as long as previous models – 15 metres compared to the original 6. The team compared a number of other designs as well but, when tested in a model tunnel, it was the train inspired by the kingfisher that ended up being faster, quieter and more powerful, with 30 per cent less air resistance than its predecessor. So it was problem solved, just like that. Well, not quite. There was a second issue, which called for another piece of animal-inspired innovation.

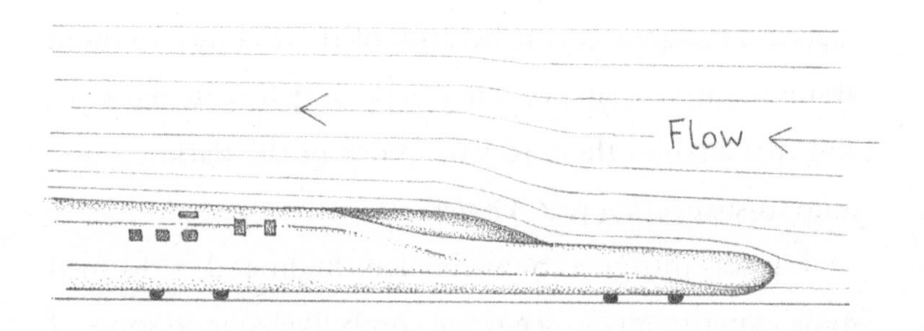

The trains obtain power from electrical wires overhead using current-collecting struts, or pantographs. You've probably seen them mounted on the roofs of trains, but not paid them much attention. Take a closer look next time you're at the station, and watch out for a train that has them mounted on top. Although they don't look like they'd have much of an effect, once a train is travelling at high speed, the flow of the air hitting them gets interrupted and creates a swirling mass of noisy miniature whirlwinds. The team tried a whole bunch of ideas, from reducing the number of pantographs to changing their shape and adding wind covers. The covers, however, were heavy; they vibrated, and the resulting noise could be heard inside the carriages. Sure, the trains were quieter on the outside, but they were now noisier on the inside.

Once again, Eiji Nakatsu found the solution in a group of birds that are known the world over as masters of silent flight. One of the reasons these birds are so extraordinary

– even a bit ghostly – is because of their ability to fly in almost complete silence. They make very little wing noise, and this enables them to swoop out of the darkness and onto unsuspecting prey. Owls!

Eiji was intrigued by their silent flight and performed some experiments to see if the comb-like edge to some of their feathers, which helps them to fly so quietly, could be applied to the bullet train. The team replaced the original pantograph design with something that looked like a set of downward-turned wings with small tabs. These tabs acted like the serrations on the owls' feathers and, sure enough, they broke up the whirlwinds and reduced the noise.

Eiji was tantalisingly close to the finish line, but there was one final obstacle to overcome: the supporting frame for the pantograph. It also needed to become more streamlined in shape to reduce the noise it made. Eiji once more turned to his feathered friends, only this time to one that's lost its ability to fly altogether: the Adélie penguin. With a body shaped like a spindle, the Adélie can move almost effortlessly through the water. After a few tests, the pantograph's supporting frame-shaft was reshaped to look more like a penguin. Lo and behold, this reduced both wind resistance and noise.

So it's not just one or two but three design problems that were solved by three different birds: three animals that really did make us smarter. On 22 March 1997, the new 500 Series Shinkansen electric train went into commercial service. Remember, the challenge here was to get passengers from Shin-Osaka to Hakata Station in Fukuoka in less than 2 hours 20 minutes, and under 75 decibels. The train came in… at 2 hours and 17 minutes and at no more than a cool and calm 75 decibels. At the time, this was a new world record for the fastest trains on Earth. Since then, bullet trains have been hurtling across Japan, leaving people and animals undisturbed while keeping the Japanese economy moving, only these days their noses are a little less bullet-shaped, thanks to Eiji Nakatsu and his riverside bird. So maybe we should call them 'kingfisher trains' instead?

2

Octopus: The Ultimate Disguise

When it comes to cool animals, there's no doubt in my mind that the octopus is a hard one to beat. It also sounds like something straight out of a *Space Invaders* game. It has three hearts, blue blood, squirts ink to deter predators, can grow back arms that have been chopped off and, being boneless, it can squeeze itself into and out of the smallest of spaces. What's most amazing of all, perhaps, is that, when it comes to trickery and deceit, these animals are the masters of disguise. They go far beyond simply changing colour. They have an extraordinary ability to shapeshift and adjust the texture of their skin, and it's this which enables them to blend in seamlessly with their surroundings. They can move pretty fast as well. When things start to hot up, they'll speed out of harm's way using a form of jet propulsion, just like a jet-ski. Once they've found a place to settle down, they simply disappear, as if by the wave of a magic wand: ingenious but also kind of creepy. Imagine if we could do the same thing, and change our clothes to match our surroundings. It would make for the coolest festival outfits, for sure, but, in a world of increasingly sophisticated surveillance and counter surveillance, the race to unlock the secrets of the octopus could have very high stakes indeed.

To better understand *how*, it's best we start with *why* they do this. Octopuses belong to the family – or more accurately the phylum – of molluscs, which means they're distantly related to animals like snails, slugs and clams; and all molluscs have one thing in common: they have a foot. Sure, it doesn't look like the foot of a human, but it does help them move around. A slug will slither along on its foot, while clams use theirs to burrow into sand or mud, but the octopus's foot has become far more specialised, evolving into a bunch of sucker-lined arms, used, amongst other things, to grab hold of fast-moving prey. Octopuses, along with squid and cuttlefish, are classed as cephalopods, which in Greek literally means 'head-foot' and describes their basic body plan perfectly: a head (or mantle) attached to a foot consisting of arms lined by suckers. Octopuses have eight grasping arms, while squid and cuttlefish have an additional two retractable tentacles that they can fling out in front of them at the last minute to more successfully catch their prey.

These characters come in all shapes and sizes, and live in almost all parts of the ocean. The prize for the longest cephalopod goes to the giant squid. When measured from the end of its body to the tip of its outstretched tentacles, it can be up to 15 metres long. The heaviest squid, though, and the largest living invertebrate on the planet, is the colossal squid from the chilly depths of the Southern Ocean. It's a touch shorter than its gigantic cousin, but weighs close to half a tonne, and regularly goes into battle with bull sperm whales, although it has to be said that those deep-diving leviathans usually get the upper hand. Even though colossal squid are massive and equipped with the biggest eyes of any living organism – each bigger than a basketball – sperm whales can still swallow them up in one gigantic gulp. At the other end of the scale is the tiny Thai bobtail squid. Fully grown, it's no more than a mighty... wait for it... one centimetre long, and weighs less than a gram.

The largest known octopus is the giant Pacific octopus, and it really is a giant. With its arms stretched out on either side, it can span up to 10 metres. Then we have the comparatively large deep-sea monster that is the seven-

Carribean reef octopus *Octopus briareus*

armed octopus, named for the way the males keep one arm that's used in egg fertilisation (the hectocotylus) safely tucked away. It has a 3.5-metre gelatinous body and feeds on jellyfish. But I have to say the most enchanting of this group of animals has to be the Dumbo octopus. A collective name used to describe a number of species, these are a group of deep-sea octopuses with large fins that would instantly remind any onlooker of the ears of Disney's most famous elephant. No matter their size, however, what's universal amongst cephalopods is the fact that they're an exceptionally bright bunch.

Octopuses, squid and cuttlefish are at the top of the invertebrate IQ league tables. Cuttlefish, for example, have the highest brain-to-body size ratios of all invertebrates, and are considered by marine biologists to be some of the most intelligent soft-bodied animals in the sea. In marine lab tests, scientists have seen first-hand how octopuses use trial and error to solve complex puzzles, and market squid are thought to have a sophisticated communication system based on changes to the colour in their skin. It is this ability of cephalopods to quickly change their appearance that draws our attention here.

So how fast can they change colour? Well, the giant Pacific octopus can do it in less than a tenth of a second to disappear before your very eyes. It's a master of disguise, although some octopuses take this ability to the extreme. The mimic octopus, for example can change its skin colour, stretch out its arms and, before you know it, you're staring into the face of a venomous lionfish, or a seething mass of deadly sea snakes. It's a brilliantly clever way of spooking off potential predators. And then there's the infamous and extremely venomous blue-ringed octopus. You can find them nestled away in tide pools in the Pacific and Indian Oceans, from Japan all the way down to Australia. When these small octopuses feel threatened, luminous blue rings start to appear all over their bodies – and there's no mistaking what this means – a flashing warning signal that says, 'Stay away, I'm venomous, so jog on, mate!'

OK, so here's the six-million-dollar question: how do octopuses change their colour? The answer lies just below the surface of their skin in thousands of colour-changing cells called chromatophores. It's these cells that are responsible for their amazing transformations.

Imagine you're holding a balloon of coloured dye. As you gently squeeze the balloon, the colour is pushed up towards the tip, stretching the balloon's surface thinner and making the colour of the dye appear brighter. Chromatophores work in a similar way. The centre of each cell contains a sac filled with black, brown, orange, red or yellow pigment. Expanding or contracting these cells moves the pigment closer or further away from the surface of the skin and, in the blink of an eye, the octopus changes colour. As well as chromatophores, some cephalopods have other colour-changing cells, which create iridescent greens, blues, silver and gold, with some even reflecting back the colours of their surrounding environment, making the animal less conspicuous.

Octopuses, as we discovered earlier, can also change the texture of their skin. It's amazing to see them in action. As they move across the ocean floor, they can literally morph their bodies to match different types of seabed, from smooth rock to spiky coral and then sand. They do this by carefully adjusting the size of projections on their skin called papillae, which results in anything from small bumps to large lumpy spikes. Suffice to say, don't even think of playing hide and seek with these guys. You will lose.

As you can imagine, all this colour-changing trickery has attracted quite a lot of interest in surveillance technology. Our first example of this comes from a collaboration between researchers from the universities of Houston and Illinois. They were inspired by the colour-changing octopus to develop a flexible skin that reads its environment and mimics its surroundings. It's made from something called thermochromatic materials. These are materials that respond to variations in temperature by changing colour. Like the very first TV sets, the team's prototypes work in black and white, creating different shades of grey, but their hope is to develop one that works in glorious technicolour. They're starting off small, with a 'skin' only a couple of square centimetres in size, although this could easily be scaled up. It's made of ultrathin layers, containing a variety of sensors, reflectors and colour-changing materials, which work together so that, like our octopus, the material changes to match the colour of the background.

I like to think of this skin as a Michelin-starred sandwich. The top layer contains a temperature-sensitive dye that appears black at low temperatures and clear at 48°C and above. The second layer is made up of white reflective silver tiles, and this is followed by an ultra-thin layer of silicon circuitry that controls the dye's temperature. The bottom layer forms a foundation of transparent silicone rubber. All together, our 'sandwich' has a thickness of less than 200 microns, about the same thickness as two sheets of ordinary writing paper. Underneath all this is an additional base layer that contains a whole load of light-sensing gadgets, and this is how the skin knows what colour to change to and when. It can do this in just one to two seconds, so not quite as fast as the real thing, but considering octopuses have had a 296-million-year head start, it's pretty good going.

Like many pieces of new technology, a skin like this may well find itself used in applications for military operations, which, if you're anything like me, undoubtedly draws mixed feelings. As a documentary filmmaker, though, what I am excited about is the thought of using camouflaged robots to film animals in their natural habitat. Not only could you get closer than ever before to observe new unseen behaviours, but you'd also avoid having an expensive high-end robot turning into an expensive mess. If you've ever seen behind-the-scenes clips of what happens to remote-filming camera buggies when big cats decide to play with them, you'll know exactly what I'm talking about. Crunch! Crunch! Crunch!

Well, maybe there's a way we could get around this? A different team, this time from Harvard University, has created a soft colour-changing robot. Like the octopus of our oceans, the robot has a soft rubbery body, so it can crawl and bend under obstacles, and can also blend in with its surroundings; it's an awesome piece of tech. Led by George M. Whitesides, the researchers designed their original soft robot back in 2011. It was made from silicon-based materials and moved by air pumped through tiny cylinders into its four arms. The robot's design has since had a bit of an upgrade, and it can now disguise itself. A network of tiny channels, much like our blood system, runs through the skin-like layer covering the robot. As different dyes are pumped in, the robot quickly changes colour. Using fluids that are hot or cold can also initiate the robot's thermal camouflage, making it extra stealthy, even against infrared night-vision cameras. If you want to do the complete opposite and, let's say, make it act as a rescue beacon, it's easy: you just pump in fluorescent fluids and, voilà, you've got yourself a glow-in-the-dark rescue bot. At the moment, these fluids are still drawn in from a separate reservoir, but this could be incorporated into the robot's body in the future.

There's still some way to go before many of these ideas are fully exploited commercially, but I'm sure you'll agree it's a very exciting area of technological innovation, and one we should probably keep an eye on... that is, if we can!

OCTOPUSES AND TRANSPLANTS

Octopuses, as their name suggests, have eight arms and these are perfect for gripping rocks, capturing prey and walking along the seafloor. Their undersides are lined with suction cups or suckers. Some species, like the curled or lesser octopus, have one row, while most others, like the common octopus, have two. Exactly how many varies from species to species. The giant Pacific octopus, for example, has arms that stretch to 6 metres, each covered in 250 suckers, making for a grand total of 2,000 suckers, the largest of which can be 6.4 centimetres across and support a whopping 16 kilograms.

Each of these suckers can be moved independently, thanks to a complex bundle of neurons and nerve cells that we call a ganglion. Ganglia are crucial when it comes to directing its touch, its sense of smell, and for the manipulation of objects. Octopuses can pull apart the shells of clams, and they've been seen digging dens and even using stones and broken rocks to create walls in order to protect the entrances to their homes. It's this dexterity, or more precisely the action of their suckers, that's attracted the attention of scientists at the University of Illinois Urbana-Champaign. Led by Hyunjoon Kong, Professor of Chemical and Biomolecular Engineering, they've been designing a device inspired by the octopus's suckers that could change the medical world of tissue transplants for ever.

Tissue transplantation is something you've probably heard of before. It's the process in which organs and cell tissues are moved from one part of the body to another; either within the same person or between a donor and a recipient. These transplantations can literally save lives and restore essential bodily functions. Let's say we have a patient with corneal blindness, for example. By removing all or part of the damaged cornea – that's the protective, transparent front part of our eyes – and grafting on a new cornea from a deceased donor, we can fully restore a person's sight.

Over the last few decades, sheets of tissue cells have been used more and more in the treatment of injured or diseased tissues. Now, when I think of the challenges of these types of transplants, one of the major things that comes to mind is tissue rejection – this is where the body's immune system sees incoming donor tissue as an invader and 'rejects' it. Way before we even get to this stage, though, an important aspect of these kinds of surgeries is one that's so basic, it barely comes to mind, and that's carrying it and getting it in place. How do you physically hold soft transplant tissue, without contaminating or accidentally ripping it?

Handling these living materials remains a real challenge, because they're fragile and they can easily crumple when they're picked up from the substance on which they're being grown. At the moment, tissue sheets grow on a temperature-sensitive soft polymer substance, which, once transferred, shrinks and releases the thin film of tissue. The only problem is that it can take anywhere between 30 and 60 minutes to transfer a single sheet, and it runs the risk of tearing or wrinkling the film, compromising the success of the operation. This is exactly the problem the Illinois-based team wanted to solve: how to quickly pick up and release thin, delicate sheets of cells without damaging them. They turned to the arms and suckers of the octopus for inspiration.

This is where we need to get up close and personal, to see how changes in pressure in the suction cups attach and detach the suckers from different objects. I want you to picture an octopus sucker: imagine a sand-timer made of rubber. If you slice off the very top, you'll be left with the shape of an open, curvy vase. Turn this vase upside down and, hey presto, you've got yourself the perfect shape and structure of a sucker, with two chambers, one above the other. The open chamber at the bottom does the sticking and is called the infundibulum, which means 'funnel,' and the chamber above is called the acetabulum, which means 'vinegar cup.' We'll call them the funnel and the cup, for short.

The funnel is quite flexible and has a ridge around its open rim. The surface is covered in a series of ridges and grooves. The cup on the other hand is more rigid, and has smooth walls. The suckers themselves contain various groups of muscles: radial muscles, which travel across the walls; circular muscles that circle around the widest part of the suckers; and meridionial muscles, which lie at right angles to both the circular and radial muscles.

When one of the suckers presses against a surface, it forms a light seal. The radial muscles then contract, and this causes the walls of the chambers to get thinner, which means the volume of the cup expands as a result. Since the funnel's sealed against a surface, water can't get in, so the same amount of water is now residing in a larger chamber. This leads to a drop in pressure and generates suction. The suckers release their grip by either relaxing the radial muscles, or contracting the circular muscles, which reduces the volume of the cup and, as a result, the suction effect.

There's also another feature that makes the suckers even more effective. They're equipped with a kind of piston or pump-like structure inside them. When something tries to pull away from the sucker, it lifts the piston, which decreases the pressure inside even further, and strengthens its grip, a bit like those Chinese finger-traps. If you haven't seen one before, it's a small, stitched bamboo tube-like toy, which, by the way, is perfect for playing pranks on your friends who are none the wiser. You invite your unsuspecting pals to put an index finger in either end. When they then try to pull their fingers out, the tube gets tighter, and the more they struggle, the harder it becomes to get out of the trap.

The Illinois team worked with researchers at other institutes in the United States as well as Chung-Ang University in South Korea. It was watching the way an octopus can pick up both wet and dry objects of all shapes, using the small pressure changes in its muscle-powered suction cups – instead of using sticky adhesives like we do currently – that inspired an idea.

They got to work making what they call a manipulator. It's roughly the size of your hand, and is essentially a suction cup attached to a rod-shaped handle. The suction cup itself looks very different to what you might imagine. When I first heard about this, I pictured something that looked more like a toilet plunger. It's in fact made up of a flat, flexible heater, and something called a hydrogel.

Hydrogels are thick jellies that are really good at holding water. In this case, the team created a gel that reacts rapidly to changes in temperature, controlled by the heater. It works like this: the electric heater is turned on, and the hydrogel heats up. Using the handle, aka the manipulator, the hydrogel is pressed against a thin sheet that the team wants to pick up. At exactly the same time, the heater's turned off. As a result, the hydrogel expands and, like the octopuses' sucker, this creates suction. The sheet is then lifted up and gently placed on the intended target. The heater is switched on again, making the hydrogel shrink and release the sheet. The entire process takes about 10 seconds, which is 180 times faster than normal.

The team think the next step lies in developing the manipulator even further to include pressure sensors that would keep track of any minor wrinkling of the tissue throughout the procedure. By adjusting the suction force in real time, they'd finally have the ability to both monitor and correct for this. They've even suggested that this system could be used for transferring fine electronic implants, and, with a few further modifications, robots could be used to transport ultrathin materials autonomously. It's still a work in progress, but it's a fascinating area of medical research, and it's all thanks to an octopus's sucker.

3

Return from the Dead: The Tardigrade

Imagine an animal that can handle being taken beyond boiling point, to over plus 151°C for a full 15 minutes, and being frozen to minus 272°C for 8 hours, and then spring back to life. Not only that, it can also withstand radiation levels a thousand times higher than any known creature living on the planet. Sounds like I'm describing the powers of a new X-Men character, right? Imagine, then, if this being was real.

Well, imagine no longer! You might not have heard of them, but I'd like to introduce you to what I reckon are not only the toughest animals on Earth, but also the cutest. They go by several different names: 'water bears' to some or, more affectionately, 'moss piglets' to others. Whatever you like to call them, however, they really are the stuff of science fiction stories, with powers beyond belief that would leave you and me looking like mummified husks. They are the tardigrades.

All that talk of dry husks has reminded me of that feeling on a hot summer's day when drinking an ice-cold, thirst-quenching drink. But what if you had to go a whole day and into the late evening before having anything to drink. Most of us would be feeling a bit faint by that point. How about two days? Now, we're getting into the territory of serious dehydration. Any takers to go beyond three days? On average, that's as long as most of us could go without a drink of water before we'd die of thirst. Only three days – a 72-hour, one-way trip to meet your maker. There are organisms out there that blow us out of the water, so to speak, and can go for much longer, although, some of

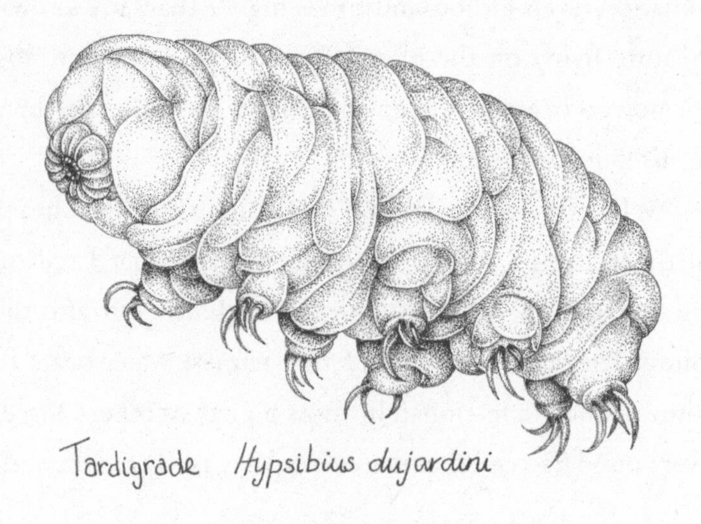

Tardigrade *Hypsibius dujardini*

their techniques might not be your first choice during a summer heatwave, or at any other time for that matter.

Let's take land tortoises of the Mojave Desert, for example. They have the ability to go a whole year without drinking water, and that's because they reabsorb it from their urine. Yeah, they're basically drinking pee. Not in the direct sense of the word, but they do retain much water in their bladder, using it like a mobile reservoir which they can draw from when they need it the most. They can reabsorb the water from their urine, so they can go for a year without drinking.

If you think that's weird, how about covering yourself head-to-toe in a slimy layer of snot while sunbathing by the pool? During long, hot, dry spells, Australia's water-holding frog surrounds itself with a cocoon of mucous, to prevent water loss. Like the tortoises, it can get water from its bladder and hold out for a couple of years until the next rains come – a neat trick if you live in the punishing central deserts of the Outback. In the Sonoran Desert, the kangaroo rat rarely drinks, and instead obtains its water from the food it eats; most vegetation, even dried seeds, will have at least some water in them. The kangaroo rat works to keep what little water it already has in its body by condensing moisture in its nasal passages as it exhales.

All these animals of the extreme can survive a shortage of water because they do their best to keep hold of it, but what if there's no water to begin with? The ability to survive almost complete dehydration is known as anhydrobiosis, which, put in everyday words that you and I can understand, means 'life without water,' and the animal that does this best is, of course, the tardigrade, which is why scientists are taking great interest in its biology. At about 0.5 to 1 millimetre in size, a light-microscope comes in handy if you want to see one clearly. I can already hear you pondering at home: 'Who has a light-microscope just lying around, Patrick?' Well, if you don't, no worries, I have the perfect recipe to help you spot your very first water bear, in five easy steps:

Step 1 Start with a nice helping of garden moss; a small clump will do just nicely.

Step 2 Next, add some of your finest local tap water, ensuring all your clump of moss is thoroughly wetted.

Step 3 Now set your timer for 20 minutes, and allow your mix to soak.

Step 4 Once your timer goes off, squeeze out all the water into a flat glass dish and let the mud settle.

Step 5 Finally, scan the dish using a magnifying glass, or, even better, get yourself one of those snazzy clip-on macro lenses for your smartphone.

If you look carefully, you should, see a whole bunch of tiny creatures moving around in the water. Keep your eyes peeled for something that looks like a miniature vacuum bag, with eight legs stuck to it. Sounds strange, but don't worry: your eyes are not deceiving you. Most of them have a pair of small black eyes, and a few even come with armoured body plates. When it comes to feeding time, some are omnivorous – in other words, they eat plants and animals, as well as algae and fungi – while others are all-out carnivores, ruthless predators that prefer to hoover up other tardigrade species. Tiny sensory hairs lining their mouths help to guide food to a specialised pharynx or tube (which bizarrely looks remarkably similar to the hole to which you'd attach your vacuum cleaner hose) which sucks food right into their gut.

All of this action is happening on a miniature scale, as tardigrades are incredibly small in size, and surprisingly cute in appearance, but the reason biologists are fascinated by them is because they're virtually indestructible. They've even survived being blasting into orbit above the Earth on a satellite. When exposed to the vacuum of space for 12 days, they miraculously came back alive. If you went into space without the protection of a spacesuit, the lack of pressure would force the air in your lungs to rush out. Gases already dissolved in your body fluids would expand, forcing your skin to inflate like a balloon, and eventually rip you apart. Your eardrums and capillaries would rupture, and your blood would start to boil. Even if you did survive all that, ionising radiation would shred the DNA in your cells. Mercifully, though, you'd be unconscious in 15 seconds, and dead shortly after. Tardigrades suffer none of those effects; in fact, about 450 of the 3,000 tardigrades that went on their space mission got lucky and produced offspring up there.

Even without human intervention, tardigrades like to get around. As well as living in the moss in your garden, they've been spotted: way up on Himalayan mountains, over 5,500 metres above sea level; in the hot springs in Japan, where the temperature can reach 100°C; at the bottom of the ocean, where the pressure is a thousand times greater than at the surface; and even on the frozen wastelands of Antarctica, where the ground temperature can be minus 94.7°C – the coldest temperature recorded anywhere on Earth. They are, you might say, the ultimate intrepid explorers and, because they can survive such extreme conditions, they've earned themselves bragging rights, and entry into a select group of animals called 'extremophiles.' Even so, they raise the question, if they're so small and live in such extreme environments, how did we discover them in the first place?

This accolade goes to Dutch microbiology pioneer Antonie van Leeuwenhoek, who first described tardigrades in a letter to London's Royal Society in 1702. Much like our tardigrade recipe, he'd taken some dry, seemingly lifeless dust from a roof gutter, added some water, and, using a rudimentary microscope (which he made himself, by the way), found what he described as 'small animalcules' swimming and crawling around. More detailed identification, though, was down to German pastor Johann August Ephraim Goeze. In 1773, he was first to refer to them as 'little water bears', because of how much they looked and walked like bears.

The name we most commonly use for them today is directly linked to Italian priest and biologist Lazzaro Spallanzani. If you haven't heard of him before, he's the very same scientist who revealed that bats navigate using echolocation. He named them *il Tardigrado*, meaning 'slow stepper', because they move so slowly. What he didn't know, but I'm sure he would've been blown away by, is that under certain conditions tardigrades slow down so much that they go into a type of suspended animation closely resembling that very final chapter of life: death.

The most remarkable thing about these creatures, though, is that, whilst they live in water, they can survive for long periods of time without it – months, years, even decades. That bring us back to our original question: how do you survive without a single drop of water?

When a tardigrade starts to dry out, it responds by retracting its head and legs, and then it more or less shuts down its metabolism to less than 0.01 per cent of the normal state. Shedding almost all the water in its body, it curls up and forms a dry husk, called a tun, which comes from the German word *Tönnchenform*. Tardigrades can remain in this protective form until they're rehydrated and spring back to life. The question is: how long can they remain in this deathly state of suspended animation? Well, the answer to that may have been stumbled upon over 70 years ago.

In 1948, Italian zoologist Tina Franceschi claimed that tardigrades could be reanimated from dried moss samples that were over 120 years old. That particular experiment has yet to be replicated but, in 1995, dried tardigrade samples were brought back to life after eight years, while, in 2017 (and every year since), a team led by Mark Blaxter at the University of Edinburgh has successfully observed live tardigrades from specimens of dried algae, which they rehydrate and check on an annual basis.

For most animals, life without water is impossible, and that's why these tardigrade resurrections have not only challenged our understanding of the boundary between life and death, but also inspired a new wave of potentially life-saving applications.

Beside tardigrades, there are a number of nematode worm, yeasts and bacteria that can also survive desiccation (the removal of moisture – think desiccated coconut flakes), and they all share a particular trait: they have a massive sweet tooth. OK, maybe not quite in the same way that we do, but they do make a lot of one type of sugar, called trehalose. It forms a glass-like state inside their cells that stabilises proteins, cell membranes and other key components, which would otherwise be destroyed.

This vital ingredient, along with the role it plays, was recognised in the 1970s by Californian researcher John H. Crowe, now Professor Crowe of the University of California, Davis. He discovered that, when the tardigrades dries up, it uses trehalose to replace water molecules in its cells. By doing this, the cells' structure is maintained on the molecular level, until water becomes available again, and the cells rehydrate once more. Professor Crowe realised that the medical implications of this discovery were truly phenomenal. If trehalose could stabilise the cells of the tardigrade, could it also protect human cells, such as blood, from becoming damaged when they dry out?

Now, of course, you wouldn't want to dry your blood while it was still inside your body – that definitely wouldn't end well – but what about blood outside the body? According to the World Health Organization, more than 100 million blood donations are collected every year. If needed, the donated blood can be separated into its different components, such as plasma and platelets, to be used in life-saving transfusions, replacing blood lost during surgery or, in some cases, to treat illnesses where patients are unable to make their own blood properly.

Platelets are particularly useful when it comes to wound healing, but they can be challenging to use and store, and here's why. You might think that you could just keep platelets refrigerated, but this would destroy them; they have to be kept at room temperature. What I was surprised to learn was they also only have a three to five-day shelf life before they become unusable and have to be disposed of. But what if you could add something to stabilise the platelets? Something like trehalose? We could then have them freeze-dried and kept as a powder for up to two years. This would give us the option of transporting blood long distances by land or air to remote locations that previously would have been impossible to reach with the platelets still intact. This new 'tardigrade-tech' is currently going through clinical trials and, if successful, blood platelets will become far more readily available for saving lives around the world.

There's also another potential application, which is sure to hit much closer to home: vaccines. There's no doubt that, since the very first treatment over 200 years ago in 1796, vaccines have improved our lives and protected us against diseases such as hepatitis B, cholera, polio, tetanus and now Covid-19. The vaccines for these diseases are hard to transport. In fact, half of all vaccines become inactive during transport, because unlike platelets, they *do* need refrigeration, which isn't always easy if electricity is unreliable or unavailable. Even if they do reach their destination intact, they still have a limited shelf life and need to be kept cold continuously. If we could dry them out using trehalose or a similar compound, vaccines could be transported and stored at room temperature, then activated by adding water when needed. In this stable, easy-to-transport form, the hope would be to extend the shelf life of these substances from mere days to months or even years, increasing access to medicines to all corners of the globe.

As well as helping out here on Earth, this tardigrade technology could be useful when it comes to space travel. Prolonged space flight is known to cause detrimental effects on the human body. American extremophile researcher Thomas Boothby is interested in how tardigrades cope with the effects of low gravity and radiation exposure. He's been leading a project with NASA to grow tardigrades on the International Space Station. His hope is that tardigrades can help us combat the stresses of space flight. If they can, then we might be able to use this knowledge and apply similar tricks to protect astronauts on lengthy space missions.

So that's how an animal few have ever seen and with a tenacity for survival is inspiring some innovative and life-saving medical applications. Not bad for a creature that looks like a vacuum bag!

4

A Woodpecker and a Black Box

Have you ever had something heavy fall and hit you on the head, or maybe had a door swing out and smack you in the face? How about the opposing team bashing into you while playing a sport like rugby? Hurts, doesn't it? The human body is a wonderfully resilient thing, but we're only able to cope with so much force when it comes to impacts before we start to bruise, get concussed and break bones or worse. But, as the world speeds up, these kinds of impacts will only get bigger and, when they come, we'll need increasingly better ways to deal with them. Luckily help is on the way – a revolutionary technology looking to deal with these high-speed impacts. We're talking about everything from improved designs for bike helmets, all the way through to the preservation of 'black box' flight recorders, which store that all-important data for solving and preventing plane crashes;

and it's all thanks to the head-banging antics of an iconic woodland bird.

Woodpeckers! These noisy knockers belong to a family of birds with such charmingly inventive names: piculets, wrynecks and sapsuckers to name a few. They're fairly wide-ranging in size. South American piculets are at the smaller end, with some no longer than your thumb, whilst the largest of the woodpeckers can be the length of your forearm. Most are forest and woodland birds, but there are a few exceptions, like the Gila woodpecker, which nests and feeds on wild saguaro cacti, native only to the arid landscapes of Arizona, California and Mexico.

Woodpeckers, as their name suggests, spend much of their waking hours clinging vertically against the side of a trees, pecking at wood. They rapidly hammer the bark in search of food like insects and grubs. They can also drill deeper holes into dead or dying trees, where the wood is slightly softer. These form little 'caves' hollowed out inside the trees, where they create their nests. And, get this, woodpeckers will use their bills not only to peck, but also to drum, communicating with other woodpeckers, to stake their claim to territories and to attract a mate. To get the best and loudest sound, they need something with really good resonant qualities. They'll drum on trees, wooden telegraph poles, and even on metal signage. If you hear them drumming, instead of pecking, you'll know they're not looking for food or drilling a hole, but are, in fact, making a very bold statement: 'This is my turf!' So how have they managed to become such superstar head-bangers? To answer that, it seems fair to start with the part of the body that's making all that racket, the bill.

As you'd expect, woodpeckers have a strong, chisel-shaped bill, a fantastic wood-drilling tool. They also have a very long sticky tongue to easily extract insects and other food from the bark. Their stiff tail and grasping toes help to secure the birds on the vertical trunks of trees. But the most interesting aspect of their behaviour (and anatomy) is how quickly they can bash their head against trees without sustaining any damage – up to 22 times a second, which is ridiculously fast. Imagine the mess, if we tried to do something similar. To understand what's going on in detail, we need to get to grips with a measurement of acceleration called G-force, the force we feel when we move forward or slow down really fast.

If you've ever flown in a plane, you'll know that, as it takes off and increases in speed, you get that feeling of being pressed back into your seat. This happens because your body's subjected to an increasing change in G-force as the plane accelerates up to take-off speed. Likewise, if you have to suddenly slam on the brakes of your car, you'll feel this same G-force as you rapidly decelerate, except this time in the opposite direction. This is very much the case for impacts too: crashing into something, or having something crash into you, are essentially very sudden moments of deceleration.

G-force, helpfully for us, is called G. These Gs increase by a factor of measurement based on Earth's gravity – 1G is equal to normal gravity, with 2G being twice the force, 3G is three times, and so on. The more Gs there are, the stronger the force, and our bodies can only take so much of it. Which makes the thrill-seeker in me wonder, what's the absolute limit? The force felt when a plane takes off would be under 2G; a very fast rollercoaster, though, would max out, albeit briefly, at around 5G or 6G; great fun, and roughly what a Formula 1 driver would experience when they hit the brakes to make a tight corner. It can get quite complicated, depending on where the force is applied and for how long, but overall, as humans, we tend to pass out if we experience 6G for a sustained period of time. There have been examples of people experiencing sudden impacts to the head of about 80G before getting concussed, but this is nothing compared to what a woodpecker's head experiences as it drums on a tree: a deceleration of around 1,200G. So how does it do it?

In order to withstand a force over 1,000 times that of gravity, a woodpecker's skull is designed to absorb shock and minimise damage. The skull can compress and expand, a bit like a sponge. The bone that surrounds the brain is packed with trabeculae (which are like microscopic plates), making it thick and spongy. This forms a tightly woven 'mesh' that provides support and protection, and stops low frequency vibrations from passing through. It's essentially armour for the brain.

In addition to this armour, woodpeckers have the equivalent of a brain 'safety belt.' In humans, the hyoid bone serves as an anchor for our tongue and helps us to swallow. It's tucked away at the front of the neck, between the lower jaw and the voice box, and shaped like a horseshoe. In woodpeckers, this bone is highly adapted. Forming a bony, springy support, it's much longer than ours and forms a loop around the entire skull. Unlike our brain – which has fissures and folds that are surrounded by fluid – the woodpecker's brain is small, smooth, and held in a very tight space to stop it from moving around too much. However, thanks to this elaborate setup, each time a woodpecker's brain collides with its skull, the force is spread out over a larger area, and it's this that makes the bird far more resistant to concussions.

A woodpecker's bill also helps to prevent trauma. It's extremely strong and doesn't bend or fracture easily. The shape is also quite intriguing: the visible outer layer of the upper bill appears to be longer than the lower bill, creating a kind of overbite, but this is misleading to some degree. If you had X-ray vision, you'd notice that the bone structure supporting the lower beak is, in fact, longer and stronger than the upper beak. Thanks to this unevenness, any impact is diverted away from the brain, and, instead, redistributed to the lower beak and bottom parts of the skull.

So the woodpecker has four different kinds of shock absorber: a hard but elastic bill; an area of spongy bone in the skull itself; a springy tongue-supporting hyoid bone that extends behind its skull; and the skull itself, which is designed to suppress vibration.

Now, the art of head-banging might not seem like the realm of a lab-coat-wearing technician, but woodpeckers and their shock-absorbing capabilities have indeed drawn considerable scientific interest. Sang-Hee Yoon and Sungmin Park of the University of California, Berkeley, became inspired by the woodpecker's skull after looking to build better structures that protect electronic equipment, such as aircraft flight recorders, from the damage caused by powerful impacts. They closely studied high-speed video of woodpeckers in action, together with CT scans of the head and neck of the golden-fronted woodpecker – a North American species – to identify the areas that absorb mechanical shock. Utilising this knowledge, they managed to fabricate a newly designed shock-absorbing system that protects microelectronics to a far great degree than previous systems.

It's made up of a cylindrical steel enclosure, a first line of defence that mimics the beak. Inside they added a layer of rubber, which works like the hyoid, followed by a second layer of metal – this time aluminium. The spongy bone was effectively replaced here by glass beads, in which the sensitive electronics sit. Yoon and Park then placed their system inside a bullet and used an airgun to fire it at an aluminium metal wall. Their results produced some astounding numbers. In their tests, the electronics were protected against impacts of up to 60,000G. Today's flight recorders can only withstand about 1,000G, which is still quite an achievement, but with an increase of 60 times, their design has much to offer, from aircraft safety to protecting spacecraft from collisions with fast-moving space debris or incoming micrometeorites.

Back here on Earth, engineers at Cranfield University, UK, who specialise in automotive impact technology, see huge potential applications in the design of regenerative shock absorbers for everyday vehicles on the road, by redirecting energy into a form more easily recoverable than the heat that's normally released. Woodpecker studies might also feed into the world of high-end motorsports, like Formula 1, where the challenge is to continually develop technologies that protect drivers during an incident in such a way as to avoid severe internal trauma. I remember

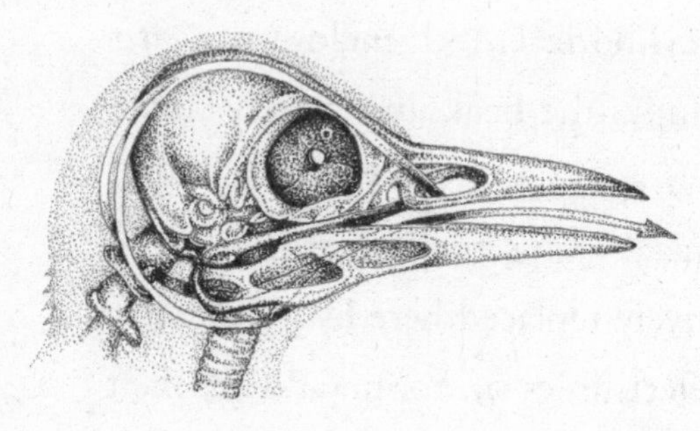

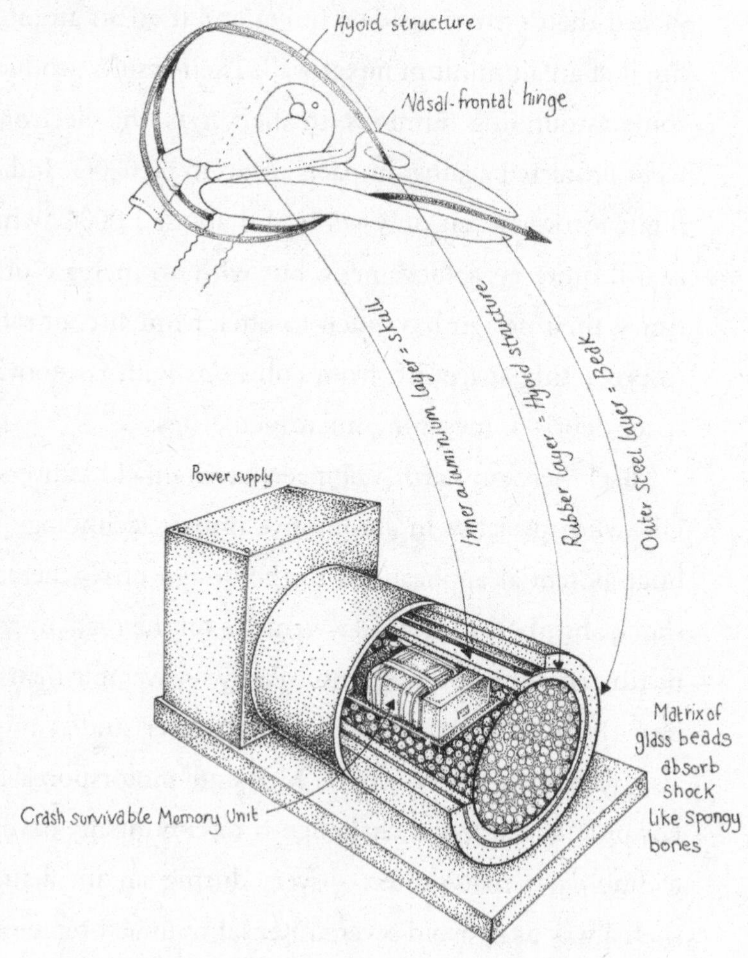
Hyoid structure
Nasal-frontal hinge
Inner aluminium layer = Skull
Rubber layer = Hyoid structure
Outer steel layer = Beak
Power supply
Matrix of
glass beads
absorb
shock
like spongy
bones
Crash survivable Memory Unit

watching Fernando Alonso's baffling crash in Melbourne in 2016. His car made contact with another at 190 mph, sending him barrel rolling through the air and into the barriers. He experienced three high-G decelerations at 45G, 46G and 20G – starter, main and dessert – a G-force full course. Still not impressed? How about Romain Grosjean's horrifying accident at the 2020 Bahrain Grand Prix? The slightest of touches with another car sent his hurtling into the barriers at top speed, in a flame-ball so intense it looked like something more akin to a Michael Bay movie than a racetrack. Miraculously, in both cases, thanks to head and neck support systems, HALO safety devices and a cleverly designed cockpit, both drivers were able to walk away with their lives intact, but not all have been so fortunate. In the future, though, woodpeckers might provide a solution.

Away from the track and onto public roads, where we mere mortals drive, Anirudha Surabhi had a less dramatic, but likewise painful, accident. Riding his bike through London he had a close call with a bus. Luckily Anirudha was wearing a helmet, but it cracked in the smash, and so he was rushed to hospital with concussion. At the time, he was doing a master's degree in design at the Royal College of Art, and was looking for some motivation for his final-year project. Suddenly, he had his answer – design a better bike helmet – with the woodpecker coming to his aid.

Surabhi was curious about the shock-absorbing spongy bone and also took particular interest in the how the hyoid wrapped around and over the top of the skull to form this built-in safety belt. He began building versions of his design using different materials: glass, rubber and cardboard. After hundreds of lab tests with each material, he finally settled on cardboard. But this wasn't any old cardboard, oh no! Surabhi created a special, dual-density cardboard with an internal honeycomb structure. To construct the liner, he laser-cut ribs of this honeycomb cardboard and assembled them into an interlocking helmet-shaped lattice. The lattice was designed with more give and flexibility than the polystyrene foam liners that are more commonly used in helmets. This flex was utilised to soften incoming blows, with the air pockets inside each individual rib acting to absorb the impact as well. The goal was to offer better head protection to the person wearing the helmet, and it worked. The 'Kranium liner,' as he called it, not only performed well at absorbing force but was also light and recyclable, because the liner was effectively 90 per cent air. Surabhi has since worked with a number of commercial companies to bring the liners and helmets to market. So the next time you hear a woodpecker pecking a pole to peeve you off, just think how this bird now stops accidents that would otherwise see you off!

5

Polar Bears and Insulation

It's a long, cold winter to the north of the Arctic Circle, where the sun remains below the horizon for months on end. For the Inuit – one of several indigenous peoples who live permanently in the northern polar region – this is the time, perhaps, to gather round a fire, bask in its warm glow, and share stories of myth and legend. One of these tales is the story of the king of the Iqsinaqtuit, a race of predators that are said to strike fear into the heart of any human. They haunt the drifting ice floes, and their powerful king only takes off his magnificent cloak in the sanctity of his own ice cave, revealing that underneath he is, in fact, human. Although the Inuit hunt down this king, they also respect him for his strength and wisdom, for they know that failing to do so would bring them great misfortune. This king goes by the name Nanuq, but he's also known by other equally mysterious names: 'the ever-wandering one,' 'the one who walks on ice' or 'the great white one.' Many of us know him as the polar bear.

This 'ice bear,' as it's also known, is one of the world's most stunning creatures. With its white coat and supremely powerful body, there's no doubt it is the most majestic of all the animals that live on the ice and snow. A quick flick through its biological stats also shows that it's very much deserving of its Inuit titles. Raised up on its hind legs, an adult bear can stand at over three metres tall and weigh up to 600 kilograms, making the polar bear the largest carnivore to walk the planet. Its only natural predators – or perhaps 'adversaries' would be a more accurate way of putting it – are people.

For centuries, humans have been in awe of the great ice bear. Its official scientific name, *Ursus maritimus*, translates as 'bear of the ocean,' which is fitting, as the polar bear spends more time roaming the vast ever-changing landscape of the Arctic Ocean's sea ice than it does walking on land. Despite its regal reputation, the polar bear is a relatively 'young' species. According to evolutionary biologists, polar bears evolved from an ancestor they shared with brown bears, including grizzlies, with which they have been known to crossbreed, producing fertile hybrids known as 'pizzlies.' The approximate date of divergence between the two species is still unclear, although the oldest known fossil of what's thought to be a polar bear jaw is about 100,000 years old, so they must have evolved prior to that; according to the latest DNA analysis, it was probably within the last 500,000 years or so. More than likely this occurred during a warm interglacial period, when the climate was mild enough for brown bears to move northwards. After the next ice age took a hold, most headed back south, but some, thanks to mutations in their DNA which code for hair colour, became adapted to the harsher conditions. Those with the most suitable adaptations, namely a lighter whiter coat for camouflage when hunting seals, were more likely to survive. Those that didn't, died. It was through this process of natural selection in action that the polar bear was born.

Over the years, polar bears changed from being omnivores to near obligate carnivores, although they will eat seaweed and summer berries when seals are in short supply. They've also adapted to an environment in which temperatures often drop to below minus 40°C for several weeks. Added to that, the sea ice that's so crucial to their survival is constantly on the move, drifting on winds and ocean currents, and so polar bears sometimes travel hundreds, if not thousands, of kilometres in search of food. When you watch them in nature documentaries, the landscape is so vast it looks as if the bears are moving in slow motion, but don't be fooled. By ambling along in what appears to be a cumbersome, almost sleepy fashion, polar bears conserve more of their energy and, even at this pace, they can easily cover a distance of up to 6,000 metres in just one hour. When they do decide to pick up the pace and sprint, however, it's a completely different story – they can reach speeds of at least 18 mph.

Polar bears are also strong swimmers. They can stay in the icy Arctic waters for long periods of time, and can easily cover over 50 kilometres in a single stretch, a necessity when you need to manoeuvre yourself between ice floes. But, with the increasing shrinkage of the Arctic sea, those journeys are getting longer. In 2011, a female bear in the Beaufort Sea was tracked swimming a record-breaking 687 kilometres – which is more than the distance between London and Aberdeen – and all because of this unprecedented ice loss. Fortunately, their massive paws, which can grow to an impressive width of 30 centimetres, make perfect paddles to propel them through the water.

Their paws are also well adapted to walking on the slippery ice, the size helping to spread the bears' weight evenly and prevent them from falling through. On the bottom of each paw are small, soft bumps called papillae, which give the bears that extra level of grip. When ice becomes particularly thin and treacherous, polar bears will slide with their legs extended outwards – imagine a cute little deer, like Bambi, on ice and you've got the picture. It does sound kind of funny, but by doing this and lowering their bodies closer to the ice, they're better able to distribute their weight, and stop themselves from crashing through.

This ability to manoeuvre on shifting ice is crucial when the polar bear is hunting for its favourite meal – seals, especially ringed seals, the most numerous seals in the Arctic. Catching one, however, can be quite tricky, and so the bears have become masters of the stakeout. Patience is a virtue they've evolved to have in spades, waiting for up to an hour next to small breathing holes made by the seals. It's the human equivalent of being sat there with a line, ice fishing. Sure, it may take a while, but for the bear, the energy-rich benefit of the seal's thick blubber is well worth the wait. When a seal eventually pops up for a breath, the bear uses its explosive speed to grab onto its prey with thick, curved claws and sharp canine teeth.

Targeting seals provides the bears with all the fat and calories they need to build up their own insulating layer of fat, which can be over 11 centimetres thick; in fact, it's surprising that they can eat so much fat without this affecting their health; although, when we take a look at the polar bear's genome, the analyses reveal that genes relating to cardiovascular function and lipid metabolism have changed somewhat in polar bears. If we ate the same proportion of fat in our diet, we'd be dead within a short period of time.

Putting fat aside, the polar bear's most important adaptation to freezing temperatures is, in fact, their striking fur, yet that brilliant white that we so closely associate with these animals is a complete illusion. Polar bears are, in fact, black. That's right, they may look white and fluffy on the outside but, underneath all that fur, their skin is really dark. They aren't born like this, though. Baby bears start off pink, and their skin darkens with age. I know it sounds like I'm making all of this up on the spot, but it's true. Scientists are still unsure why, but a likely explanation could be to protect the bears from harmful UV rays, plus black is the best colour for absorbing that precious and much-needed heat energy from the sun.

But hang on a second! Isn't this absorption effect completely offset by the fact that polar bears have white fur? White reflects heat radiation, so what's going on here? Again, and this might come as an even bigger surprise, although polar bear fur looks white, it's actually close to being transparent. The bears have two different layers of fur, one on top of the other. The first is an insulating, short undercoat of dense hair next to their skin. The second is a layer made up of what's known as guard hairs. These are coarse, tapered hairs which grow to a length of about 10 centimetres out through the undercoat. Both layers of fur are pigment-free, making them translucent. The only thing that stops them from being totally see-through is the protein they're made from – keratin – and that gives them a slightly off-white appearance. The hairs appear to be bright white, however, because of the way light reflects around them. It's for this reason that polar bears look their whitest when they're surrounded by ice, and the sun is at its highest in the sky. This is especially true if you see them after their moulting period, which usually begins in May or June and ends by August.

In the lead-up to this, the oils in their fur – from eating all those seals – can make them look a bit yellow, and if you happen to catch a glimpse of these magnificent creatures during a beautiful pastel-coloured sunset, polar bears will appear to be the same colour, because this light is reflected around the animal. There's even the strange phenomenon of having *green* polar bears. In this case, though, it's not because of the reflection of green light. It comes from living in captivity. Algae from the bear's enclosure can sometime manage to grow *inside* its fur, turning the bear a pale green. It's the way the guard hairs are constructed that provides opportunity for algae to get inside the fur. You see, each hair contains multiple hollow, air-filled chambers. These are all lined up inside each hair shaft, one after the other, and are vital to how the fur works. Firstly, the hollow hairs make the fur super-lightweight, helpful when you're a big bear moving around in search of food. Secondly, and most importantly, they trap warm air, making the fur an incredible thermal insulator.

There is one situation, however, when the fur isn't as good at keeping our Arctic specialist warm, and that's when the bear's swimming. Although it is water-repellent, the fur doesn't work so well in the icy polar waters. In this situation, the bears rely more on that thick fat layer to keep warm. Walking on land and ice, though, polar bear fur is so effective that bears have been known to overheat when they run.

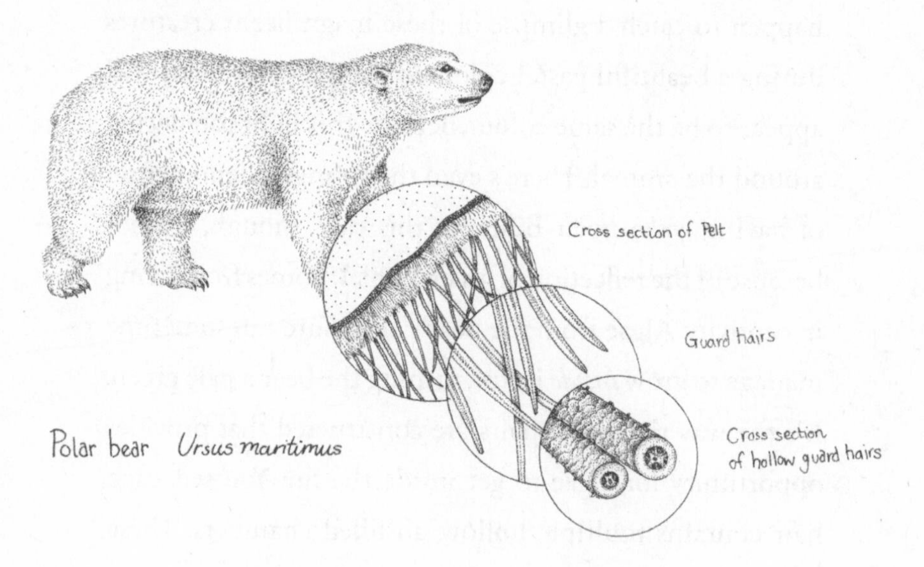

Polar bear *Ursus maritimus*

The Inuit have been aware of the fur's properties for generations, which is why they use it in their boots and clothing. In the last couple of years, it's also caught the eye of a team of scientists, who are using polar bear fur as the model for a new type of insulation for spacecraft. A team of researchers at the University of Science and Technology in China has been inspired by polar bear fur and developed a new type of aerogel that can be used in space.

OK, so what's aerogel? Well, it's one of the coolest materials I've ever heard of, and I'm not the only one who thinks so. A quick search on the internet and you'll soon discover why. Aerogel is a remarkable human-made substance, whose story began with a bet, back in 1931. American scientist and chemical engineer Samuel Stephens Kistler wanted to win this wager after one of his friends set him a challenge to remove all the liquid from a jelly-like substance, without changing its shape. I'm not exactly sure how Vegas bookmakers would calculate the odds for that one today, but, trust me, Samuel and his buddies were the cool kids of their time. After thinking about it for a while, Kistler came up with a sure way to collect. He managed to develop a method of removing liquid from a jelly by increasing both the temperature and, crucially, also the pressure around it. This process changed the liquid inside into a gas-like superfluid. The result was a jelly that kept its initial shape on the outside, while inside it had transformed into a network of fibrous material, surrounded by holes where the liquid it was holding used to be. And thus, aerogel was created.

Aerogel is often nicknamed 'frozen smoke' or 'solid smoke' because you can see right through it in the same way you can see through real smoke, except aerogel is a solid mass, plus it has some pretty mystifying properties. All those empty holes mean it has a very low density, making it extremely light – one of the lightest solids known. Even though it's brittle and breaks easily, it can support a lot of weight: 1,000 times its own, to be exact. This means a tiny two-gram piece of aerogel can support a brick weighing two kilograms; and, because it's difficult for heat to travel through the network of holes, from one side to the other, aerogel is a fantastic thermal insulator.

Despite aerogel having been around for some time now, it's only recently that scientists have started to explore all the creative ways it can be used. Take building insulation, for example: you could potentially replace frosted glass windows with aerogel. It would still allow light through, but without letting heat move in or out. Even more exciting, aerogel has now been deployed in a place that's literally out of this world. In 1997, NASA scientists used aerogel to protect the operating systems of the

Pathfinder rover from overheating or freezing on planet Mars, where temperatures can swing wildly between plus 40°C during the day and minus 40°C degrees at night.

Much closer to home, here on *terra firma*, the researchers in China, led by Professor Shu-Hong Yu, wondered whether they could create a new and even more effective type of insulating aerogel, by copying the structure of polar bear hair. To do this, they constructed millions of carbon tubes containing air-filled chambers, replicating the air-filled chambers found in the polar bear's guard hairs. Each tube was tiny, roughly the size of a single strand of human hair. The team bundled the tubes together into a single block, and this became the basis for the new aerogel unit – think of a load of spaghetti, wound up, round and round itself, all piled at the bottom of a lunchbox, and you'll have a good idea of what all these tubes look like, although this happens on a much smaller scale.

Compared to other aerogels, the team found the polar bear-inspired design was much lighter and even better at stopping heat from passing through it, and, like polar bear fur, it was water-repellent. It also had one other useful and *unexpected* quality: it was remarkably stretchy and flexible, even more so than the hair of a polar bear. This was really exciting for the team, because they realised it could be used in far more places than the traditional, more brittle version of aerogel. At the moment, the team has only made small blocks of their polar bear-inspired aerogel, but they're looking at ways to scale up the manufacturing process in order to make it in larger sheets. If they can crack this challenge – or in the case of aerogel, stop things from cracking – they're confident that their solution could be extremely useful for the aerospace industry, which means we could see more of it being used up in space.

Scientists are considering aerogel as the answer to human colonisation of Mars. It's possible that a two-to-three-centimetre-thick layer could be all that's needed to construct an atmospheric, aerogel greenhouse, which would protect the humans living inside both from extreme temperatures and dangerous UV radiation, while still letting enough light in to grow crops in support of the colony. With its lightweight properties, its ability to regulate temperature and its newfound flexibility, I wonder what the Inuit would make of King Nanuq's cloak as the key to making futuristic dreams of interplanetary colonisation a reality.

6

Mosquitoes, Wasps and Advances in Medical Technology

Like many people, but for the nauseating feeling of a sharp needle piercing my arm, I'd love to become a regular blood donor. I'm sure that's the case with loads of other people too. Which raises a question: if so many of us want to help, why don't we? What are we so afraid of? I guess it's the associated pain that turns most of us ashen when faced with a long needle. According to the World Health Organization, there are more than 16 billion 'safe' injections administered every year, which equates to nearly 44 million injections every single day. But what if all those injections could be made less painful? The secret, it seems, lies in the mouth of an insect, one that annoys us with the jabbing of its needle-like mouthparts to feed and gorge on our blood. It's gruesome to watch up close, but for a group of researchers getting under the skin of this process makes total sense. After all, the thing about this bloodsucker, is that, although you may hear it coming, you rarely feel its bite.

Mosquitoes! They're about as welcome as a bonfire on an ice rink. For a wildlife filmmaker, some of the most beautiful and biodiverse ecosystems to capture on film are jungle rainforests. It's here that you find some of the planet's most recognisable animals: sloths, monkeys, hummingbirds, and giant anaconda. However – and keep this between you and me – jungles are one of my least favourite locations to film, because they're also home to the most aggressive mozzies you'll ever meet in your life, and film crews are the ideal mobile feeding station. Trekking through thick foliage with heavy camera kit often pulls the straps of carry cases tight against your clothes, and this makes an ideal drop-zone for these airborne lancets. Personally, I just wish they'd buzz off and leave my crew and me alone in peace. Well, we can all dream, can't we?

These thoughts probably echo the feelings of sixteenth-century explorers as they moved through the rainforests of the Americas. In fact, the word *mosquito* appears to have originated in post-Columbian North America, in the

1580s. It comes from the Spanish for 'little fly,' and that's exactly what they are – a type of fly in the order *Diptera*. Many of us are familiar with their appearance: like all true flies, they have one pair of wings, a slender segmented body, feathery antennae, three pairs of hair-like legs, and famously elongated mouthparts. Now it wouldn't feel right to talk about mosquitoes and not mention one of their greatest onscreen appearances. *Jurassic Park* is one of the most influential movies of our time, but I've always wondered, could any of what we saw have been rooted in real science? Well, yes and no. As of this moment, we're yet to have found, let alone successfully extract, any remnants of dinosaur DNA from fossilised remains (that would be so cool, though) – it's just too old. However, it is true that ancient mosquitoes, similar to the ones of today, have been found encased in amber that's 79 million years old. Since dinosaurs were around between roughly 243 million and 66 million years ago, this is a great indication that mosquitoes of that time (like the one encased in amber) probably fed on the blood of dinosaurs.

Today, there are more than 3,000 species of the insect, which are found pretty much in every region across the globe (except Antarctica, Iceland and a few islands in the Indo-Pacific region). Wherever there are mosquitoes, we're sure to have mosquito-borne diseases – malaria, yellow fever, dengue fever and West Nile virus, to name just a few. Most of these diseases are found in the tropics. However, it may come as a surprise to learn that malaria was once prevalent as far north as England, where (carried by the gnat genus *Culex*) it was known as 'intermittent fever' or – in brilliantly Dickensian fashion – 'the ague', and so there is some concern that, with a warming global climate, it could return once again.

Now, for all their haemophilic antics, not all mosquitoes are blood feeders. While it's difficult to be sure, a rough guess is that fewer than 14 per cent of mosquito species feed on human blood. Even so, it's mosquitoes – not sharks, lions or even venomous snakes – that hold the unenviable title as the 'world's deadliest animal', and that's because they are the perfect vectors for spreading viruses and other disease-causing microorganisms, which result in the death of millions of people worldwide. The World Health Organization found that malaria alone, which is transmitted by the *Anopheles* mosquito, was responsible for an estimated 445,000 fatalities in 2016. Dengue fever, zika, chikungunya and yellow fever are all transmitted to humans by the *Aedes aegypti* mosquito. With more than half the world's population living in areas where these mosquitos are present, you can understand why this little fly has the potential to cause such big problems.

Why, though, do they even need our blood in the first place? When it comes to the main culprit, it's the female who's the bloodsucker here. Male mosquitos live the easy life and feed on the nectar and honeydew of plants, but the female needs to work much harder and track down a blood meal before she can lay her eggs. It's the proteins within the blood, needed for egg development, which she's after. And, like all good mothers, she knows exactly where to direct her attention in order to give her young the best start in life. She begins by seeking signs invisible to the naked eye: carbon dioxide from our breath, body heat and a number of volatile fatty acids emanating from our skin. A clear indication that she's out hunting is the notorious high-pitched whine she makes after sneaking into your room at night. You switch the light back on, and the whine seems to stop. Switch it off again and that annoying whine starts up once more. It's so annoying and feels like an endless war of attrition until, in the dark, that moment in the dark when the sound stops abruptly – and you know she's about to feed.

Once she's landed, she searches for an area of skin where the blood vessels are close to the surface and then sets to work, but it's only after she's punctured your skin, fed on your blood and left the scene of the crime that the itching begins and you notice a bump appear. The itch is

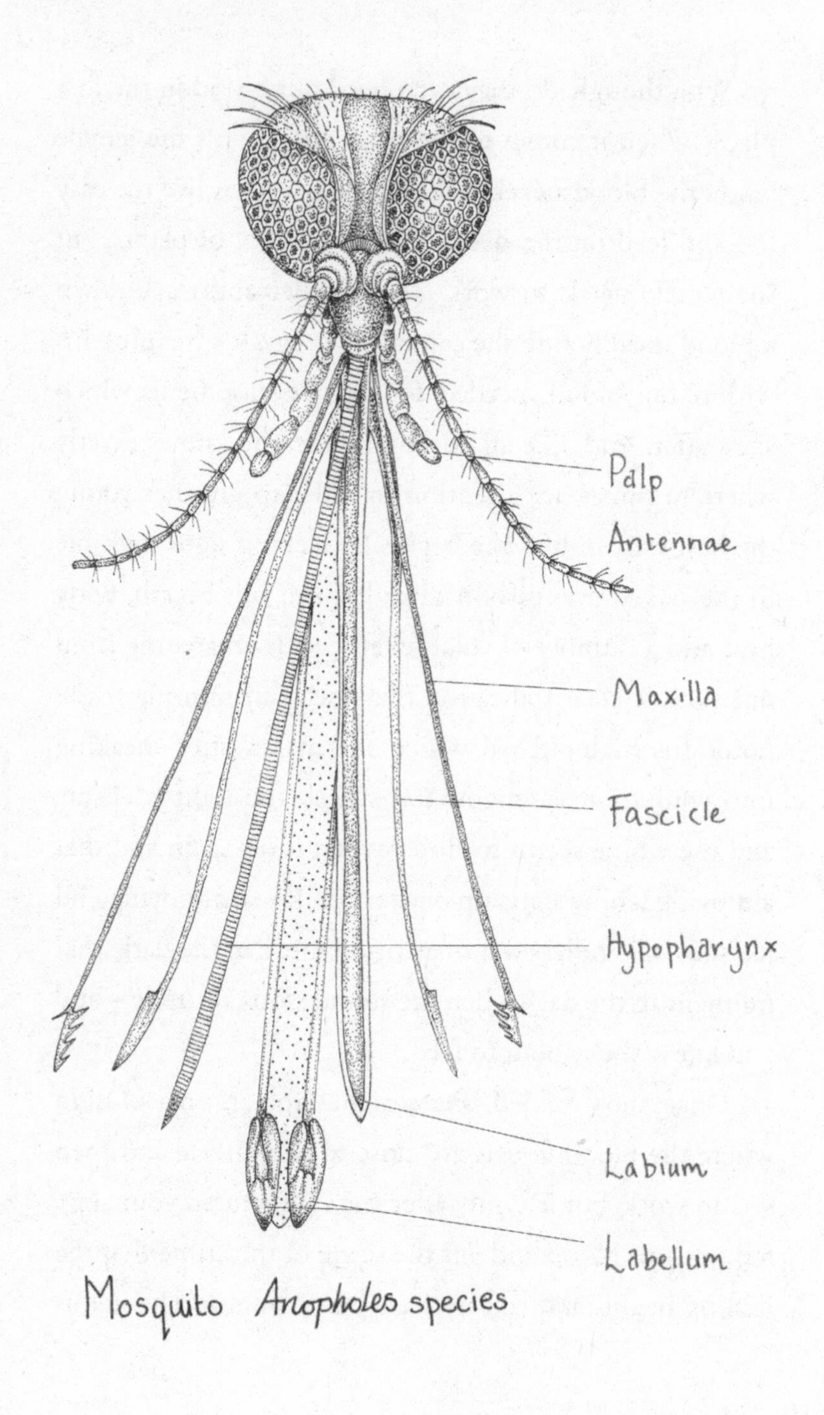

Mosquito *Anopholes* species

nothing to do with the bite itself. This is caused by bacteria in the anti-coagulant injected by the insect to prevent your blood from clotting. When mosquitoes are doing their worst – penetrating your skin and sucking your blood – you don't feel a thing, and it's all thanks to the design of their mouthparts.

Mosquito mouthparts are incredibly complex, and are made up of several stylets (or probes) inside a retractable cover, which form a long proboscis. Let's start with the labium, a retractable sheath that covers the other mouthparts. When a female mosquito lands, she uses this to gently press against the surface of your skin, before the other mouthparts are inserted. Next up are the mandibles and maxillae, which do the actual piercing. Mandibles have pointed ends and go deeper into the skin. Maxillae end in jagged blades, which grip the flesh as they penetrate the host… that's you, by the way. The mosquito even vibrates its head to work all these parts more easily into the flesh. Once in, these are driven even deeper, so, inside the host we now have two tubes: one pumps saliva down into the flesh, which numbs the skin (and explains why we don't feel anything to begin with); the other sucks up the blood. Some experiments have seen mosquitoes feeding for three to four minutes without their host feeling a thing, and so scientists are looking to the mosquito's marvellous mouth to design a pain-free surgical needle and provide humanity with the answer to the painless injections of the future.

Seiji Aoyagi and his colleagues at Kansai University, Japan, have developed a needle that mimics the mosquito's proboscis. So how does this differ from needles that doctors and nurses normally use? Traditional medical syringes, as we know, have a sharp end to pierce through the skin, and a smooth surface along its length that enables it to pass through skin and flesh uninterrupted. But a lot of the pain you feel when it's inserted isn't really because of the pointy end; it comes down to the fact that all that metal along its length comes into direct contact with your skin. In contrast to this smooth needle, the mosquito's piercing stylet is serrated.

You might think that this would make it even more painful; I certainly did. What happens, though, is the complete opposite. The serrations make very little contact with the skin, which reduces stimulation of pain receptor nerves, and, as a result, you feel less pain. Using high-speed video microscopes, the team of scientists observed the sequence of events as the mosquito fed and, eventually, they were able to replicate it themselves.

Taking inspiration from the mosquito, they created a needle etched from silicon, that was 1 millimetre long and 0.1 millimetres in diameter; that's about the thickness of human hair. It had two harpoon-like, jagged-edged outer shanks, which penetrated the skin first, after which a drug-delivering and blood-sucking tube moved down between them, only touching the patient at the point they exited the sharpened tip. And, just like the mosquito, which vibrates its proboscis to help the maxillae ease down through the tissue, Seiji's design also mimicked this movement, with each of the three parts of his device vibrated by tiny motors.

Their intricate design also had to be durable. In the early stages of development, the needles were very brittle, which, of course, would cause major issues if they broke inside a patient. They developed a method of testing their strength, by pushing the needle into a piece of silicone rubber which was wrapped around a tube containing red dye. They watched as a tank attached to the needle slowly filled with the dye, confirmation that the needle was, indeed, capable of successfully puncturing skin without breaking. Then came the human trial. Like a true scientist, Seiji tested the needle on himself and a number of volunteers. They all confirmed that the pain was far less than expected, although the period of injection lasted longer than it would've done with a regular syringe. The new needle was also fitted with a small tank to store the blood or fluids that it collected. This would enable doctors to analyse the sample using a fibre-optic cable connected to the tank.

The team hope that their needle will pave the way for a range of small, wireless, injecting or blood-collecting devices, permanently attached to the body. They might end up injecting you several times a day with that life-saving medication you need, or collect samples to monitor blood sugar levels if you have diabetes, or to monitor blood cholesterol if you're at risk of heart failure. Only this time, just like the mosquito, you wouldn't feel a thing.

TAIL-END TOOLS

While the female mosquito has a probe at the front end of her body, the female parasitoid wasp has an equally sharp probe at the other end. We're all familiar with the black-and-yellow-striped wasps (known as yellowjackets in the USA) that invade our open-air picnics and threaten us with the stings in their tails, but the parasitoid wasps are even more sinister. Some of them lay their eggs deep inside the bodies of other insects. When the larvae eventually hatch, they start eating their hosts from the inside out. Yuk! Whilst this might be the stuff of nightmares for most of us, it's got scientists very excited, because it's these wasps, that jab their egg-laying tube into their unsuspecting hosts, which may hold the answer to vastly improving keyhole surgery techniques, and, in the process, save countless lives.

Let me introduce you to *Glyptapanteles*, a genus of wasps found in New Zealand and also in North and Central America. When the female wasp is ready to lay her eggs, she searches for a suitable host. Like the female mosquito, this parasitoid identifies her target partly by smell, and in this case the chosen victim is the caterpillar of the geometrid moth *Thyrinteina leucoceraea.* Having tracked down the caterpillar, the wasp drills into its flesh with a long thin tube called an ovipositor, which literally means 'egg placer' – a needle-like organ with which she stabs her host. She then proceeds to lays her eggs inside. Much to my horror, Arne Janssen, an ecologist at the University of Amsterdam in the Netherlands, says that 'it can pump as many as 80 eggs into the caterpillar.' When the eggs hatch into larvae, they begin feeding on the caterpillar's body juices, taking care to avoid any vital organs – at least these wasps have some table manners – and then go through several moults, whereby they shed their skin as they grow. The caterpillar, meanwhile, seeming unperturbed by its attackers, continues to feed and grow.

Eventually, the wasp larvae are ready to make their exit, and they all emerge together. What's even more extraordinary is that, when the larvae emerge, the caterpillar is still alive. It's thought that the wasp larvae time their final moult to coincide with their exit, so that as they squeeze through the caterpillar's skin, the moulted skin which they leave behind blocks up the exit hole. So, they perform their own simple surgery on their seriously injured host. And, just when you think this story can't get any stranger, there's one last play to run.

Not all the larvae leave the caterpillar. A few of them stay inside and their job is to control the now 'zombie caterpillar' and make it guard the escapees. Meanwhile, the emergent larvae turn into the next stage of the life cycle, as pupae, which look like small, pale brown, pods, or seed cases. The caterpillar takes up position near the pupae, arches its back, then stops moving and even stops feeding. It will, on occasion, spin silk over the pupae to protect them, but the mind-boggling part takes place when it's disturbed: it begins to thrash about violently, defending the wasp pupae from potential predators, such as shield bugs. When the pupae hatch, the new wasps emerge and fly off, and the caterpillar finally dies. If all of this is making the hairs on the back of your neck stand up, I'm right there with you. Having said that, I think it's time to turn the tables and give these parasitoids a little credit, because it's that egg-laying needle that's inspired the design of a fascinating new tool for keyhole surgery.

The extraction of tissues from the human body is an important part of many different kinds of surgery. This might be to examine abnormal tissue or lesions inside the body, or to remove tumorous, dying and infected tissue from a patient. Removing tissue on or near the surface of the body is, for the most part, straightforward, but extracting tissue from deeper inside the body is far more challenging. Various methods have been developed, and some of the most effective are what's called aspiration-based devices. The term 'aspirate' means to suck or inhale. So aspiration devices use a sucking action to remove tissue or other targeted items from an area inside the body.

There can be problems with this technique. Sometimes these devices become blocked, when removing things like blood clots, for instance, or the tissue could be damaged as it's removed, causing problems when it's examined later. There's also the possibility that healthy tissue could accidentally get pulled into the tube. Even with all the scientific breakthroughs and our current level of medical technology, it's not easy to reach remote locations in the human body, in tight spaces or on tiny structures, for example, inside the brain, because of the relatively large size of the instruments. Aspiration-based devices are currently one of the best options out there, but there may be an improved alternative on the way.

A team at Delft University of Technology in the Netherlands has developed a new device, inspired by the ovipositor of parasitic wasps. Noticing that the ovipositors were so thin, and thus had no space for muscles inside, the researchers, led by Aimée Sakes, knew there must be a clever mechanism at work. If they could find out what it was, and how it worked, then they might be able to recreate it for life-saving keyhole surgery.

The ovipositor is shaped like a flexible, hollow needle. Precisely how it works is still being studied, but there are three valves inside, joined by a tongue and groove mechanism. This arrangement is a nifty way of fitting two surfaces together. Imagine you have two flat pieces of wood. Each piece has a slot along one edge, called the groove, and a thin deep ridge on the opposite edge, called the tongue. The tongue of one piece can fit inside the groove of the next piece and this keeps them together, and this is how the valves inside our tube fit together. It's a little complicated, but let's stick with it.

There's one large valve, and two smaller valves. To understand how they fit together, imagine a pipe that you slice lengthwise into three pieces: a large piece and two smaller pieces. The three pieces slot back together and slide back and forth along their length using the tongue and groove method. Inside the tube, the valves can slide independently of each other to create friction forces. To lay an egg, one of the smaller pieces (or valves) slides deeper into the body. Then the second small valve slides forward, followed by the large valve sliding forward, using the two others as a sliding support. This motion of the valves is repeated several times over and the friction generated is used to move the valves and the tube deeper inside the body. This motion is thought to also be responsible for moving the egg along inside the ovipositor tube.

Inspired by this design, a prototype made up from a series of needles has been designed at the Bio-Inspired Technology Group at Delft University. The needles have four to six equal-sized rods that can be individually controlled. In the experiment, the same principle of friction forces used by the parasitoid wasp was used to propel the needle through a substance made of gelatine. Based on the results of these experiments, the next step for the team was to turn the system inside out. Instead of using friction to propel the needle through a substance, they used friction to propel a substance through the needle.

Their transportation system consists of six blades held together by an outer tube. The tube is about the width of a matchstick. The semi-cylindrical or curved blades are equal in size and shape. The tube and the blades are attached to another piece of kit, which sets the blades in motion and enables them to slide alongside each other. Let's imagine the tube is being used to remove a piece of cancer tissue. The blades travel in the direction of the cancerous cells one at a time. As one blade moves forward and cuts the tissue, five slide backwards, removing the cancer cells as they do so. The fiction caused by having one blade move in the opposite direction to the other five, pulls the material away and up through the tube; in fact, depending on the movement of the blades, tissue can be transported in both directions.

Whilst the prototype has great potential, the team have identified improvements which need to be made to both its design and operation before clinical trials can begin. But what the prototype *has* demonstrated is that friction-based transport has the potential to become a viable and reliable alternative to existing methods – and this is exactly what they set out to explore. The scientists believe this system could eventually become more precise than the standard suction tubes used during keyhole surgery – and potentially enable the removal of tumorous tissues deep inside the human body through tiny incisions. Aimée Sakes hopes to see clinics using this device as soon as 2025. So, you never know: if, at some future date, you find yourself having keyhole surgery, you might just have the egg-laying, caterpillar zombie making parasitoid wasp to thank.

7

Master-Builders: Termites

Termite mounds are extraordinary things. Like small dormant volcanoes with additional side chimneys and funnels, they look so alien in the landscape. The termites build the mound from clay or mud, and live in a nest at its base. One of my first encounters with them was in Namibia, where bat-eared foxes seek them out. The fox is one of the smaller carnivorous desert mammals, and gets its common name from its disproportionately large, bat-like ears, which it uses like satellite dishes to pick up the sounds of its dinner. And termites are one of its favourite snacks.

We don't often think of termites as being noisy creatures so, to find out how the bat-eared fox finds its grain-sized prey, I lowered a tiny microphone down one of the holes of the termite mound. When I put the headphones over my ears, what I heard was something from the world of the weird and bizarre. It sounded like I'd tuned in to a radio station dedicated to popping candy, that same sound you get when you pour milk over a bowl of crisped rice cereal – crackle, crackle, pop, snap, pop! It made me realise that, whilst everything seemed to be calm and sedentary on the outside, inside the mound, the termites were really busy at work. My presence didn't go unnoticed either, and eventually a few termites came out to explore, and started blocking up the hole I was using with fresh muddy saliva.

Back home in London, hanging out with some friends, I was reminded of those termites. It was a hot summer's day, and the temperature of the retail store we were in was starting to rise. Surrounded by crowds and crowds of people, with the exit nowhere in sight, I started to feel trapped. Even writing about it now is getting me a little hot under the collar. You know how it is: you slowly feel yourself getting warmer and warmer, as the air continues to get thicker and thicker. There seems no escaping the stuffy store you're in. You can't open any windows, and there's nowhere close to enough air conditioning. Beads of sweat start forming on your brow of your forehead, and you can't wait to get out, get home and into the shower. So how can we keep shopping centres well ventilated and their customers nice and cool? It appears that that the building habits of a blind subterranean insect have the answer.

The science of heating and ventilating buildings is a challenge that architects and designers regularly face, architects like Mick Pearce who was hired in 1991 to design the largest office and retail building in the Zimbabwe capital of Harare: the Eastgate Centre. He had a major financial hurdle to overcome, though – the cost of keeping the building cool. He had to come up with an efficient cooling system for the building, which, unlike existing options, was cheap to run.

For Mick, this was an interesting challenge because he describes himself as someone who chooses to work within three parameters: 'aesthetics, resources, and nature'. It's that last one we focus on in this chapter. By nature, he was referring to the Gaia theory of natural systems, in which the Earth acts like a living system, and living organisms interact with their surroundings as part of a self-regulating system that supports life. Basically, it all comes down to living in harmony with the planet. He believes that designers need to see our cities more like ecosystems, in which all parts are connected and influence each other. With this in mind, he wondered, would it be possible to design a building that regulates heating and cooling all by itself? Then, one day, the solution came to him, whilst sat watching a BBC wildlife documentary... about termites.

The termites in question live in a very hot climate. In an effort to stay cool, they live in nests below the ground, at the foot of huge mounds made of sand, clay or mud mixed with saliva. These mounds can tower over nine metres high; that's higher than four human beings stood on top of each other. With such a large colony, there's a real danger that several million insects living together in such proximity would cause the air inside a mound to very quickly become stale and anoxic (low in oxygen), but the termites have solved that problem too. It was then that Mick realised the solution to his problem was staring him right in the face.

Scattered across the forests and grasslands of Africa, Asia, South America and Australia are societies of expert builders. They've been described as looking like 'grains of rice with big heads and hedge trimmers for mouthparts,' which doesn't necessarily inspire confidence in urban planning. These tower-building species of termites, however, know exactly what they're doing when it comes to engineering their homes; and they need to because they can only survive if the colony stays within one degree of 31°C. The temperature in places like Zimbabwe can fluctuate as much as 20 degrees between night and day, so how do the termites manage to regulate the airflow and temperature inside their mounds? What's their secret?

Well, it turns out, the large mound of soil that each colony constructs above the nest acts as a natural ventilation and air-conditioning system.

For decades, scientists have marvelled at these towering mounds and how they worked. Although it was widely believed that these insect superstructures helped with ventilation in the nest, exchanging stale air for fresh, how this was achieved remained a mystery. But scientist Hunter King and his colleagues at Harvard University were able to put forward a solution to this mystery with the help of a fungus-growing species called *Odontotermes obesus*, which lives in tropical southwest Asia, together with thermal imaging and the installation of tiny air-flow sensors in about two dozen termite mounds.

Their studies revealed that each mound acts like an 'external lung' and uses the change in temperature between night and day to drive ventilation. It's all based on the principle that hot air rises and cold air falls. Another way of thinking about this is that termites build towers that optimise the mixing and exchange of gases like oxygen and carbon dioxide, much like our lungs. The mounds, therefore, are an extension of their metabolism – a living system – a superorganism, if you will.

It works because, far from being a solid structure, the walls of termite mounds are covered in tiny pores, invisible to the human eye, which allow air to pass through. There's also a large central chimney, connected to a system of pipes located in the mound's thin, flutelike side mounds, called buttresses. During the day, the air in the thin buttresses warms more quickly than the air in the insulated central chimney. As a result, hot air rises through the outer chimneys and cool air in the central chimney sinks, so the air circulates, continuously bringing in oxygen and flushing out carbon dioxide. At night, the ventilation system reverses: the air in the outer mounds or buttresses cools quickly, falling to a temperature below that of the central chimney. This reversal in air flow, in turn, expels the carbon-dioxide-rich air that builds up in the subterranean nest during the day, the result of millions of living bodies respiring and using up oxygen. As an extra touch, some species of termite will constantly adjust the mound, alternatively opening up new tunnels and blocking others to more actively regulate the heat, humidity and overall ventilation inside... It also stops annoying film crews from shoving microphones down their tunnels!

Meanwhile, back in front of his television, Mick Pearce realised he could use what he was seeing in the termite mound to ventilate the shopping centre he was designing in Harare, the influences of which you can see to this day. The site is made up of two buildings, linked together by a glass roof. A network of walkways and steel bridges spans the atrium below, with lifts and escalators between the various levels and the sky walkways. The buildings are made from concrete slabs and bricks, and, just like the soil inside our termite mound, these materials have a high thermal mass, which means they can absorb a lot of heat energy without changing in temperature very much. The exterior of the building is prickly like a cactus, which increases the surface area of the building. This has the effect of boosting heat loss at night, whilst heat gain is reduced during the day.

Inside the buildings, electric fans suck up cool night air from outside, and blow it upstairs through hollow spaces under the floors, and from here into each office through vents. As the air rises, it's drawn out through 48 round brick funnels. On cool summer nights, fans are used to circulate air through the building seven times an hour to chill the hollow floors. During the day, fans circulate two changes of air per hour through the building. In this way, the air is kept fresh.

At night, the concrete blocks are cooled, and this chills the circulating air. When morning comes and the temperature rises, warm air is vented up through the ceiling and released by the chimneys. Thanks to this design the temperature is maintained at comfortable levels; and,

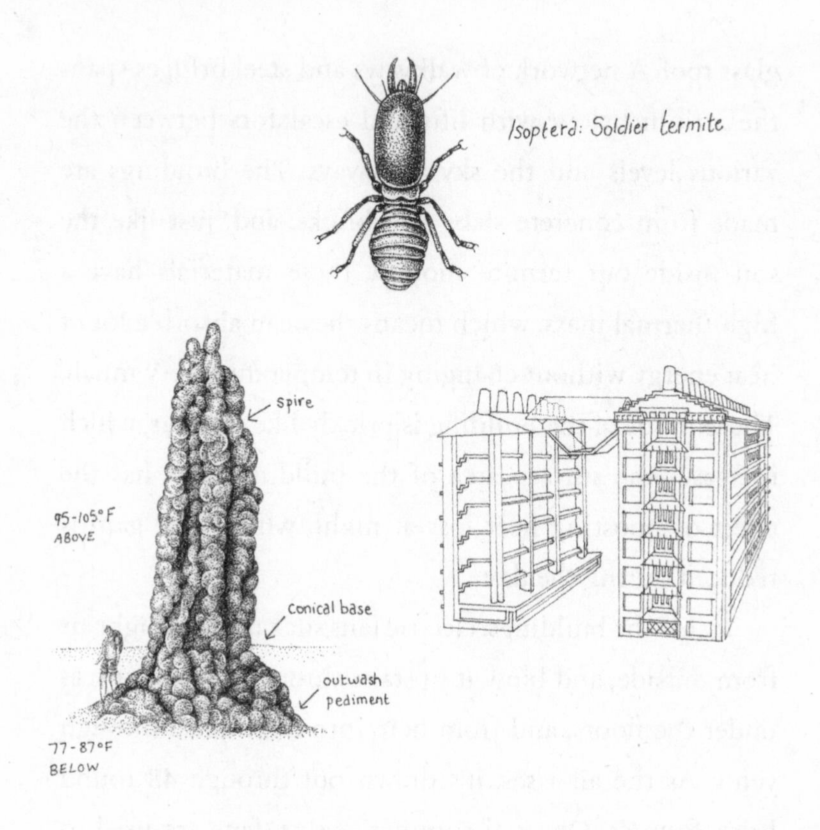
Isoptera: Soldier termite
spire
95-105°F
ABOVE
conical base
outwash
pediment
77-87°F
BELOW

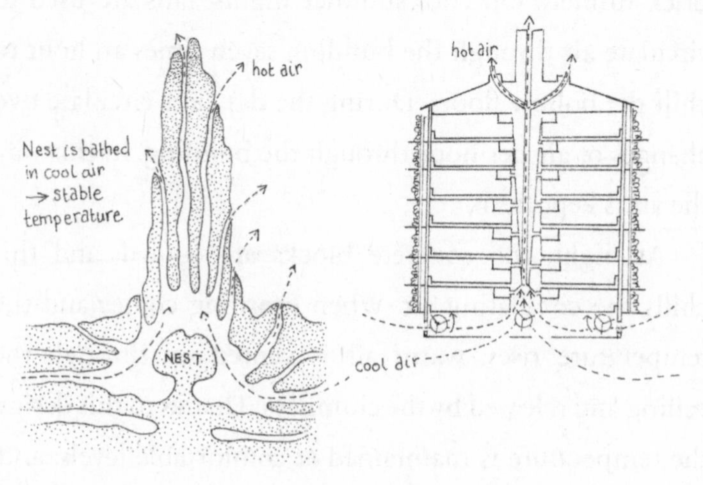
hot air
Nest is bathed
in cool air
→ stable
temperature
NEST
hot air
cool air

while there's still room for improvement in this award-winning design, it uses 50 per cent less energy than buildings cooled with conventional air conditioners.

Like the termites and their finishing touches, Mick went one step further, working with a team in Australia to design and construct the Melbourne City Council House 2 building, also known as CH2. The face of the building is made up entirely of vertical timber slats, which cover a fully glazed wall. These slats pivot vertically, like the doors in your house, opening and closing in response to the time of day and the angle of the sun. It's almost as if the structure comes to life, as it moves in response to the conditions around it. By doing this, CH2 maintains an internal temperature of 21–23°C, and uses 80 per cent less energy than buildings of the same size.

Someone else who's long been fascinated by termite mounds and their ventilation systems is Rupert Soar, an engineer at Nottingham Trent University, UK. Taking lessons from the termite mound, Rupert believes that we should try to create buildings that are permeable and allow for better air flow. He argues that buildings are generally designed as hermetically sealed boxes, meaning they're virtually airtight. He believes this needs to change, with walls becoming more like membranes. As well as saving energy, through the harnessing of natural air flow, instead of one powered by electricity, this 'membrane mentality' would also eliminate problems like dampness. Of course, it would require a fundamental shift in the way buildings are conceived and produced. But, who knows – in years to come, we might see the end of traditional bricks and mortar, to be replaced by more lightweight breathable structures – just like the mounds of termites, quietly keeping their cool, all while living in the hot desert.

8

Cod and the Cold

I have a confession to make: I'm a bit of a sun worshipper. OK, you got me there, I lied…I'm a massive sun worshipper! You know what I'm talking about, that feeling of the sun's warming glow on your skin, charging you up with every single ray. Everything just feels better when the sun's out. Everyone around smiles more and, in the summer, it feels like life is just one big holiday. But, as winter approaches and the days begin to darken and the temperature starts to drop, I'm not at my best. While scenes of snow-covered fields and mountains are very beautiful, in towns and cities it's more like everything turns to a grey slush. I used to dread those early winter mornings, when I had to scrape the frost from the windscreen of my mum's car. I'd much prefer to be somewhere more Mediterranean, a pleasant 24°C with a fresh sea breeze. This is the reason I have the utmost admiration for the plants and animals out there with the ability to live at temperatures that wouldn't just leave our teeth chattering, but would also freeze human blood. Fish, like cod, and invertebrates, like snow fleas, for example, can survive these sub-zero temperatures without freezing, and they have become the inspiration for new technologies that could extend the preservation time of transplant organs, and enable us to fly them even further, and that could make all the difference when it comes to saving someone's life.

The problem with living at very low temperatures, though, isn't necessarily the cold itself, but what happens to the water within the bodies of these living organisms. As the temperature falls below freezing, small ice crystals begin to form inside the cells of plants and animals; this is perhaps more apparent in creatures that can't regulate their own body temperature – the so-called cold-blooded animals. When these ice crystals grow, they draw water out of the surrounding cells, which destroys their structure, and this ultimately kills the cells. If you've ever tried to cool down a bottle of juice or wine in the freezer, but left it for too long, you'll know all too well what happens. As it freezes the water inside expands, and, eventually, the liquid bursts through, leaving you with an icy mess and one refreshing drink down the drain. Foods like berries turn to mush after being left to thaw out for exactly the same reason – it's because their cells burst once they've been frozen.

So, what actually happens when water freezes? As water gets colder, it becomes denser, all the way until it reaches about 4°C. By the time it reaches freezing point and turns to ice, however, the arrangement of the newly formed ice crystals means the water molecules now take up more space, and so it all becomes lighter and less dense. This is why ice floats on top of water. The expansion is also what causes materials to crack and rupture, like a burst water pipe during winter. This can be damaging to plant and animal cells too. It's the expansion of the water as it freezes that causes them to burst.

However, some living organisms contain a group of unique molecules called 'antifreeze proteins' or AFPs, although the term 'ice-binding protein' has been proposed as an alternative name, because AFPs don't necessarily stop freezing as such. They do, however, stop cell damage, when things get a little too chilly. These AFPs have a remarkable ability to lower the freezing point of a solution, which keeps the ice crystals very small and thus more manageable for the organism in question. Fish that live in the Arctic and Southern oceans are fantastic examples of this. They can survive in temperatures of around minus 2°C, because they have AFPs in their blood. These lower the freezing point of the water in their bodies, which enables them to inhabit icy cold seas below the freezing point of their blood serum, which is minus 0.7°C. Living in these conditions, most other fish would die.

These species avoid freezing through a mechanism called 'adsorption inhibition.' When the antifreeze molecules bind to an ice crystal, they change its flat growth faces into regions with curved faces. The water molecules can't easily bond to these and so they remain as a liquid rather than being converted into solid ice. So the only way to get the water molecules to stay in these curved regions is to lower the temperature, which means the antifreeze molecules have, in effect, lowered the freezing point.

It's not just polar fish, though, that possess this sub-zero superpower. A whole host of organisms – bacteria, fungi, crustaceans, micro-algae, and amphibians – have also developed strategies to survive freezing conditions. But let's stick with fish for now.

Take the Antarctic notothenioids of the Southern Ocean, for example, which include the aptly named crocodile icefish (which has no haemoglobin in its blood, so looks ghostly white), as well as the Atlantic and Arctic cod species of the Arctic Ocean. They all live in seas that are really cold, and really salty, and it's this combination of cold and salt that makes these seas quite a difficult place to live.

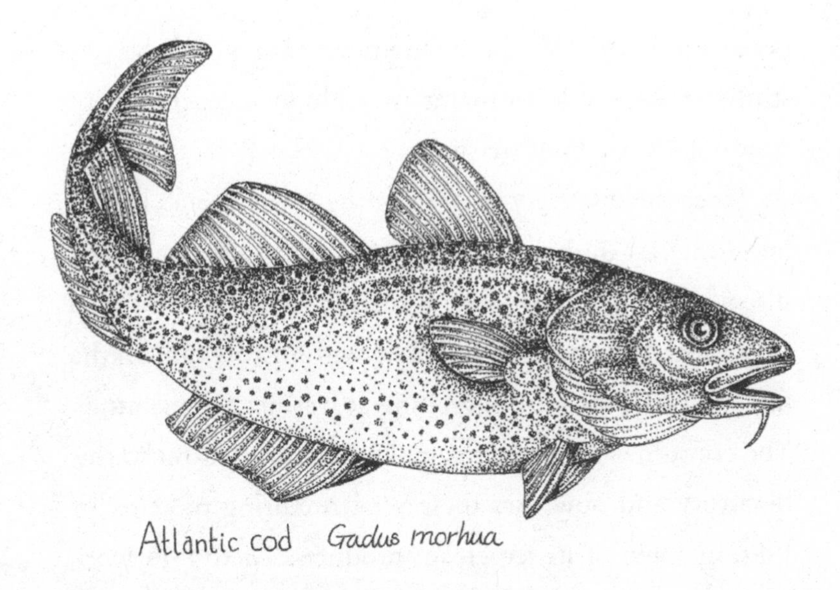
Atlantic cod *Gadus morhua*

There's also something even weirder going on here. We have freezing cold seawater, which still hasn't frozen. This is because seawater doesn't freeze at 0°C, like freshwater would. Because of the salt dissolved in it, the freezing point of seawater drops from zero to minus 1.9°C. This very cold seawater – which is a liquid – should instantly freeze the less salty water inside a fish's body, but it doesn't. So, how's the fish getting around this?

The physiologist Professor Arthur DeVries, of the University of Illinois, revealed it's the action of sugary proteins, called glycoproteins, attaching themselves to ice crystals in the blood of Antarctic fish, which prevents the ice crystals from growing. This, combined with naturally occurring body salts, meant the notothenioid fish he was studying were able to maintain a blood temperature of minus 2.5°C without freezing.

Since the discovery of AFPs in the 1960s, there's been a lot of interest in their economic potential. One fascinating discovery is that the same antifreeze proteins that keep organisms from freezing in cold environments can, on the flipside, prevent ice from melting at warmer temperatures. The consumer goods company Unilever caught on to this discovery, and now uses these 'ice-structuring proteins' or ISPs, in some of its ice cream products. These ISPs work in a similar way to the AFPs found in our Antarctic fish, although they can be produced in the lab using genetically modified baker's yeast, and so don't relying on dwindling fish stocks. The yeast contains a gene from a fish called the ocean pout, which is found in the northwest Atlantic Ocean, and has antifreeze proteins in its blood. The result is an ice cream that, as the temperature rises, maintains its shape for longer and melts much more slowly than it normally would. If this seems a bit of a niche market, it's worth noting that by 2024, global ice cream production is projected to be worth an estimated US $74 billion. But, it's not just ice cream where antifreeze proteins are proving useful.

For many airlines, keeping their aircraft frost free in cold conditions is a major challenge. Any delay on the ground due to ice on airfreight or passenger aircraft incurs a cost, as well as the potential risk of damage to the plane. The longer the delay, the more expensive it becomes. In Germany, the Fraunhofer Institute for Manufacturing Technology and Advanced Materials is currently working on a variety of strategies for minimising ice formation, including an anti-icing coating, all based on antifreeze proteins. These proteins would protect the surface of the plane from icing over and keep the planes moving – good news for the airlines and fewer delays for us as passengers. Still, I wouldn't mind taking that free upgrade instead.

Next up on the ice chopping block – power lines. Not only can the weight of ice bring down the cables, it also acts as an insulator. The ice causes the wires to do the opposite of what we think might happen, and makes them heat up. This makes energy transmission far less efficient. So, perhaps, by treating power lines with AFP coatings, there could potentially a saving of millions of dollars in maintenance and repairs costs alone, adding to the benefit of improved energy efficiency.

But perhaps the most exciting technology that's being developed from this field of study is within the health sector, where AFPs could have a critical role in preserving organs and tissues destined for transplant surgery. With thousands of people waiting for a transplant worldwide, the ability to transport organs further afield or even storing organs for longer periods of time would be hugely beneficial. One of the reasons why this might be possible is because scientists are now able to produce new synthetic AFPs, and it's all thanks to the humble snow flea.

Snow fleas aren't fleas at all, and the name is given to several different arthropods, but it is one in particular – a springtail – in which we have an antifreeze interest, especially a dark blue North American snow flea, which looks like flecks of black pepper on the snow.

Springtails are a primitive group of organisms, which are closely related to insects. They're wingless, with simple eyes, which detect light and dark only, and unusual mouthparts inside, rather than outside, their head. Like insects, they have six legs and move over the snow by walking or jumping. Instead of using their legs to jump, like true fleas, they catapult themselves into the air by releasing a spring-like mechanism called a furcula, that's folded under the abdomen. When the furcula's released, the snow fleas are launched up to 30 centimetres into the air, which is mighty impressive, seeing as these little creatures are less than three millimetres long. It's a jump of 100 times its own body length – that's the human equivalent of jumping over London's St Paul's Cathedral with 60 metres to spare. They live mainly in soil and amongst

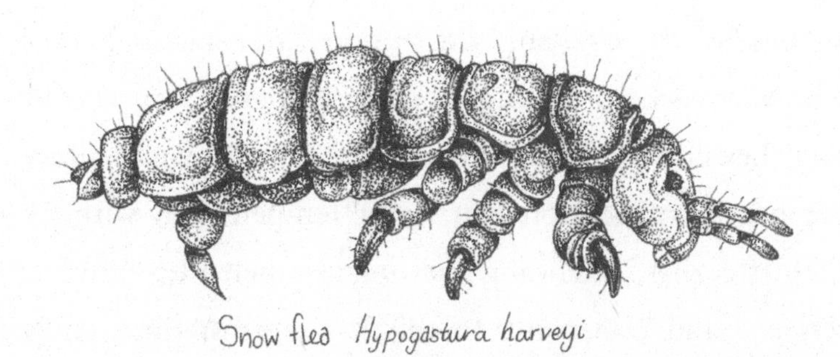
Snow flea *Hypogastura harveyi*

leaf litter, feeding on decaying bits of fungi, algae, bacteria and other organic matter. As their common name suggests, when most other insects (and sensible humans, like my mum) are cosily tucked away, snow fleas can survive in sub-zero temperatures, and are often seen jumping about on patches of snow.

Professor Laurie Graham from Queens University, Canada, first noticed them whilst cross-country skiing, and collected several to take back to her lab. It turns out that the snow flea is protected from the effects of freezing by AFPs rich in an amino acid (a protein-building block) called glycine. Just like in the cod, this binds to ice crystals and prevents them from growing larger, and so the water in a snow flea's cells doesn't freeze. By isolating the protein, the scientists were able to study its structure more closely and develop new synthetic AFPs, which, amongst other things, could be used in the transport of organs for transplant surgery.

One of the challenges with organ transplants is that you want to keep the organs super cold, but not so cold that they end up frozen and damaged. AFPs would allow the organs to be stored at lower temperatures without them freezing, extending the organs' 'shelf life' between removal and transplant. In this race against time, AFPs could make all the difference in saving the life of some fighting for survival. An added benefit of these proteins is that, at higher temperatures, AFPs lose their structure, so they'd degrade naturally once up to body temperature, and then be cleared quickly from the patient's system, reducing the possibility of any side effects.

The research into antifreeze proteins is still in its infancy, but already scientists are recognising a wide range of potential applications. What we don't know is how climate change might affect many of the species that benefit from AFPs. We can only advance the field of synthetic AFPs if we have more of these organisms that make them naturally, living and thriving in remote parts of the world (such as the Antarctic) as they have done for millions of years, prior to our arrival. It's a cautionary warning that in order to explore and benefit from this planet's biodiversity, we'll need to fight for its protection.

9

Elephant Trunks and Bionic Arms

Imagine that you've been born with a long, fat noodle attached to your face. Understandably, you have no idea how to move this giant noodle around; in fact, it would seem it has a mind of its own! Over the next few months it will, at times, prove to be a little awkward, but life without it would undoubtedly spell certain doom for you and your fellow kin. Sounds like the opening words for the memoirs of a ramen enthusiast, right? Wrong, you've just entered the world of every baby elephant to have ever walked the planet.

An elephant's trunk requires precise coordination to use correctly. Luckily, when they grow up, elephants have evolved the biggest brains of all land mammals, so to develop and strengthen the neural pathways from the brain to the rest of their body requires time. As you can imagine, a newly born elephant calf's trunk may seem more of a floppy, awkward obstacle, than a life-changing appendage. I have to admit, I found it hard to believe that elephants have to learn how to use their trunks, but that's precisely what I witnessed on a visit to the Sheldrick Wildlife Trust in Kenya, which was founded in 1977 by Dame Daphne Sheldrick. The Trust works alongside the Kenya Wildlife Service, the Kenya Forest Service and local communities to secure safe havens for wildlife.

Among many other endeavours, the project rescues, rehabilitates and reintegrates orphaned elephants into a new family herd. To help the young calves through this journey, keepers look after them and guard them round the clock, some even sleeping overnight in the stables with their precious babies. Other than sleeping, feeding is an essential part of the process of bonding with their surrogate human herd members. It was also the perfect opportunity for me to get up close and see their trunks in action. Armed with a four-litre bottle of milk, containing a special elephant milk formula that Dame Daphne was the first to perfect, the sight of what can only be described as a train of 20 elephant calves, all lined up in single file and thundering towards the feeding station, in the golden morning light, was one I'll never forget.

The thing I found most surprising was their size. Standing next to the two-year-olds, which were already at shoulder height, was quite intimidating, especially as they jostled for position around the keepers and myself. In fact, one of the older calves, clearly unimpressed by my feeding skills, slowly coiled its trunk around the bottle and proceeded to confiscate it from my hands to feed itself. As they fed, the difference between the more experienced calves and those much younger was very clear to see. One of the orphans, just a few months old, was so small that it had its very own blanket that the rangers would drape over its back, to keep the little elephant warm during chilly mornings and the cool evenings. She fed at a much slower pace, and was far more hesitant when using her trunk, which, on occasion, moved so awkwardly, it looked like it was being controlled by someone else.

In the wild, it takes elephants about ten years to learn all the skills and knowledge they'll need to make it to adulthood. Using their trunk takes a whole year, at the very least, to perfect, and that's arguably the baby elephant's most important lesson. Once mastered, though, the elephant's trunk becomes its strongest, most skilful and vital tool. It's no wonder that scientists wanted to replicate this strength and dexterity to create a manoeuvrable, but soft arm that could revolutionise the world of robotics for ever.

We came across this particular story thanks to one of our *30 Animals* podcast listeners. It was seven-year-old Pranav Sanivarapu from Bangalore who wrote to us from India to suggest we look into a story he'd seen about how elephant trunks were inspiring a new robotic arm. So, Pranav, I'm pleased to say, 'Thanks for the tip-off, and this one's for you!'

There's no debating it, an elephant's trunk is a truly marvellous thing. Soft to the touch, it's reinforced with cartilage – the same stuff that human joints are made of – which provides the trunk with strength, toughness and durability. The best way I can think to describe the texture of the skin is to consider a cross somewhere between velvet and very light sandpaper. Up close, it resembles the type of ribbed hose you'd expect to find on a vacuum cleaner. The upper or forward-facing side of the trunk is covered in hundreds of lightly bristled hairs, whereas the rear-facing side is more flattened, and is lined with small bumps that run along its length. These folds and ridges of skin reminded me of miniature mountains and valleys but, once stretched out, the ridges flatten out to resemble the open plains of the savannah upon which the African elephant roams. I always thought there was something wonderfully poetic about that. The finger-like projections at the tip of the trunk are used to investigate nearby surroundings. The way it moves reminds me in some way of a leech, searching the air for the warmth of its next host – well, without the creepy bloodsucking part! Elephants are way cuter than leeches, in fact, as the biggest animals walking the planet today, I think it's only fair to give you some big numbers to show you just how awesome they truly are.

Their trunk contains a whopping 40,000 muscles. It's strong enough to push down trees and lift more than 300 kilograms in weight, but it's also sensitive. At the tip of the trunk, are the 'fingers' – African elephants have two, while the Asian elephant only has one – which enable the trunk

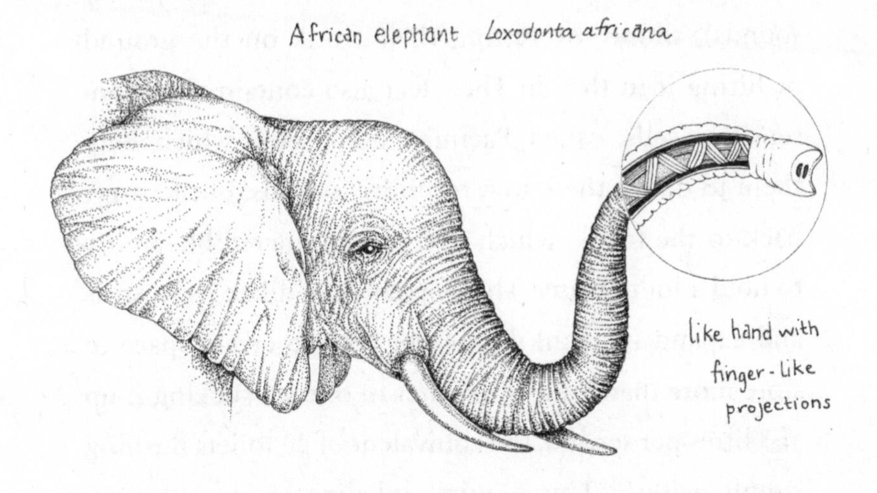

to perform delicate manoeuvres, like picking up a single blade of grass. It has two long nostrils, which guide scents to one of the best olfactory nerve centres of the entire animal kingdom. Olfaction, by the way, is the process of sensing smells, and elephants are brilliant at this, they can even sniff out water – yes, water – from a distance.

Not only is the trunk a super sniffing device – an elephant's sense of smell is thought to be four times greater than a bloodhound's – it's also sensitive to vibrations. Elephants can use it to check out the presence of distant herds, and even thunderstorms. They detect the incoming waves of infrasound (very low-frequency sounds), simply by resting their trunk on the ground or lifting it in the air. Their feet also contain vibration-sensitive cells, called Pacinian corpuscles, which help them to detect these low, rumbling sounds. But let's get back to the trunk, which also has the incredible ability to hold a lot of water. The elephant can dilate its nostrils and expand its trunk by 64 per cent to create space to store more than 9 litres of water in one go, sucking it up at 3 litres per second, the equivalent of 20 toilets flushing simultaneously. This requires inhaling air at 330 mph, which is 30 times faster than a human sneeze and comparable to a high-speed railway train. With this same trunk, the elephant can rip up trees, hoover up piles of food, and yet delicately pick up a tortilla chip from a table without breaking it.

Contrary to popular belief, though, the trunk isn't like a straw, used directly for drinking. It moves the water to the elephant's mouth, from where it can then take a drink. It also comes in handy for spraying water and a dusting of dry mud over the elephant's back and ears to help it cool down and to repel external parasites. And, when crossing deep water, it doubles up as a snorkel. It's just brilliant – an elephant's muscular multi-tool.

The trunk can be used also as an object of affection and comfort. A mother will wrap her trunk over her calf's legs and belly, whilst making soothing rumbling sounds. Elephants are even known to pet and cuddle themselves with their own trunks to make themselves feel better.

It's this versatility – its ability to grip and be strong and flexible, but with a delicate touch – that's caught the eye of engineers at the Bionic Learning Network, a research project set up by a consortium of universities, institutes and development companies, including the German industrial and automation company Festo. The reason for their interest? Well, engineers have noticed an issue with the massive robots used on manufacturing lines, like the ones you see assembling cars in factories. You might have seen videos of these robots performing their tasks to a mesmerising degree of accuracy. They're powerful, precise and almost balletic in the way they move. Although they're brilliant at performing exact tasks at exact times, if their sharp, metal surfaces were to encounter a human while in motion, the result for the human would be painful. So robots and people need to be kept well apart. This isn't always practical, though, since there are times when you would want to get close to a robot: if, for instance, it was being used in a medical procedure.

So, in 2010, the German-led team decided to develop a robot that would be far less dangerous, and yet just as flexible and strong – and that's when they started looking at all the special properties of an elephant's trunk. The engineers wanted to imitate the softness of the trunk in their robotic arm, so decided against using hard metals like steel or iron to build it; instead, they used a lightweight plastic called a polyamide. To construct the arm, they employed what was revolutionary technology at the time, and still is, if you ask me: 3D printing, a computer-controlled process by which materials can be joined or solidified into a 3D shape. Instead of the arm being a solid mass of plastic, the engineers designed it to contain lots of hollow chambers, stacked up, one on top of another. This lighter design, as the team say, makes the robot far less likely to hurt a human being, even if it were to accidentally hit someone with force. And with that, the 'Bionic Handling Assistant' was born.

Having solved the problem of creating a lightweight arm, the engineers had to work out how to make it as flexible and as strong as an elephant's trunk. They had their minds set on the fact that their robot should not contain anything hard or hazardous, so the decision was made: make it work using just air. In everyday life, air seems pretty average, but when you put it under pressure – in other words, if you compress it – it becomes a powerful way of lifting heavy loads and moving things around. The team built a series of valves and a control unit, stored at the base of the robot out of harm's way, which would work to turn air into compressed air. They used these valves to send compressed air up into the hollow chambers inside the arm.

To imagine what these chambers look like, think of lots of tiny, empty balloons inside the length of the arm. The valves are used to inflate and deflate these balloons with varying amounts of air and, in this way, move the arm around. So, for example, if the valves inflate the balloons on the left side of the arm, these stretch out and elongate, the same as if you blew up a balloon. This effectively pushes the arm in the opposite direction, in this case, to the right. And, there you have it: a robotic arm that's lightweight, can move in any direction, and is strong enough to pick up heavy loads; but, this isn't the only way the Bionic Handling Assistant mimics the elephant's trunk.

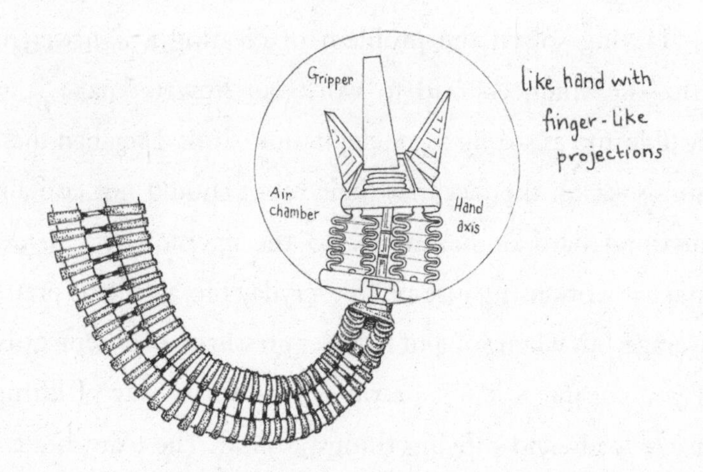

At the tip of the soft arm, the team installed fingers, just like the fingers at the tip of an elephant's real trunk. These are also inflated and deflated with compressed air to make them move. They're constructed from a soft material that can mould itself around any object, even if it's extremely fragile. This means the fingers have no problem picking up something as delicate as an egg or as thin as a blade of grass, just like the elephant can.

Back in 2010, when the Bionic Handling Assistant was first developed, its soft and flexible qualities were pretty revolutionary. Since then, it's inspired many other robots that are even more flexible. In fact, in the last few years, the field of soft robotics has exploded, and there are hundreds of different models now being developed. This means more human-friendly machines that can be used in many situations where they couldn't before. In the future, we might even get used to soft, robotic arms taking over everyday tasks we find a bit boring or dangerous – how about an automated window cleaner, or a soft robot to wash your car? One of the design team even programmed the robotic arm to pick up the toys in his children's playroom. Pretty cool, considering its humble beginnings as a sensitive, flexible noodle, hanging in the middle of an elephant's face. The elephant's trunk is truly a marvellous thing!

10

Birds, Bats and Bots

Searching the open ocean for signs of survivors from the wreckage of an air or sea disaster is a near impossible task. Ships and aircraft can only investigate relatively small areas at a time. The search area is largely determined by the coordinates of the last mayday signal but, without those, the search site can often be too vast and remote for humans to search effectively. Imagine that we could, instead, send out a team of flying search robots, each gliding effortlessly over the waves, covering vast distances, while using minimal energy. In the future, that might be possible and, for this one, we look to an animal that's perhaps nature's most astonishing feat of aerial evolution: the wandering albatross.

When I think of animals that are truly masters of the skies, the wandering albatross is a strong contender for the top spot. It's an enormous seabird, with the largest recorded wingspan of any bird, measuring up to 3.5 metres across – about the same length as a small car – plus, its wings are uniquely adapted to soar thousands of kilometres with the minimum effort.

Given that albatrosses nest on remote islands and spend several years of their adult life out at sea, many of us are unlikely to have seen one in the flesh, but, if you can picture something that looks a bit like an overgrown seagull, with a large pink bill and pink feet, then you're on the right track. For centuries, the albatross has captured the imagination of artists, poets and writers. I became more familiar with these birds after working on the BBC documentary series *Frozen Planet*, but I must preface that by saying the first time I ever heard about this group of

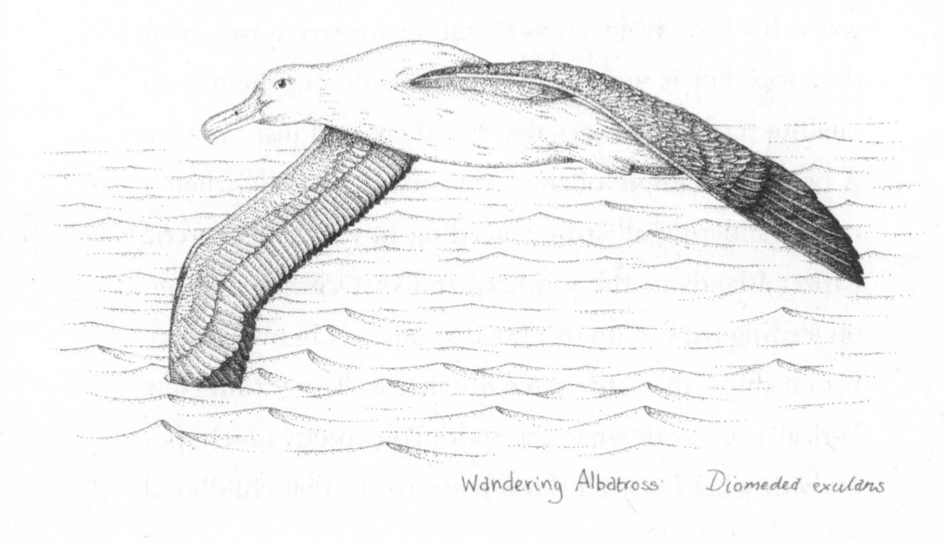

Wandering Albatross *Diomedea exulans*

birds called albatross was in a childhood movie favourite, *The Rescuers Down Under*. Although, he wasn't technically a wandering albatross – the biggest albatross of the *Diomedeidae* family – the character, Wilbur, was famed for his ability to fly our heroes across long distances, in this case New York to Australia non-stop. Away from the big screen, though, these graceful birds, have caught the attention of aerial robotics designers, as the inspiration for a new flying machine that harnesses the power of both wind and water.

Wandering albatrosses are a kind of wonder bird. They spend most of their lives not on land but out gliding over the waves, only coming back to solid ground to breed. Their range stretches right across the Southern Ocean and up towards subtropical waters – that's a lot of air to cover. It's long been known that albatrosses travel great distances, but it was only once we could tag them with satellite tracking devices that we discovered just how far. A brilliant example of this came about in 1989, when a male wandering albatross, breeding in the sub-Antarctic Crozet Islands in the southern Indian Ocean, would go on feeding trips of up to 15,000 kilometres between incubation shifts; that's the same distance as flying from New York all the way down to the South Pole (pretty much spot on with what I saw all those years ago in that childhood movie). Over the course of a single year, some individuals have managed to circumnavigate the Southern Ocean three times – that's a distance of more than 120,000 kilometres, over a third of the way to the moon.

Their efforts become even more impressive when you consider how the distance an albatross can fly would come at a huge energy cost for most other birds, meaning they'd need to eat far more food. The albatross, however, can stay aloft, skilfully gliding through the air with barely a wing beat. It's this ability that has scientists fascinated. Imagine, if we could design a machine that could fly in a similar way.

Looking back through the history books, the bird's method of flying into a prevailing wind caught the attention of the seventeenth-century merchant trader, traveller and writer Peter Mundy. He was struck by how albatrosses were able to glide so effortlessly close to the water, without flapping their wings. It set him thinking about how some ships did better when they sailed in a similar fashion. He wondered if the answer might lie in the way the wind struck the slightly curved underside of the bird's wing, and, likewise, the curved hollow formed by a ship's sail.

Some time later, in the early nineteenth century, French sea captain Jean-Marie Le Bris became fascinated by the flying skills of the albatross he saw on his travels. He went about carefully studying those that had been captured and preserved. Using them as the basis for his own model, he constructed and flew two lightweight gliders referred to as *L'Albatros artificiel*, which, as you might've guessed, is French for 'The Artificial Albatross.' Then, in the early twentieth century, American ornithologist Robert Cushman Murphy made the same connection. Watching albatross fly in a South Atlantic gale, he wrote: 'When the secret of their perfect balance has been learned and applied to man-made planes, then we'll go aflying.'

The albatross's secret is all down to something called dynamic soaring, in which the bird harnesses the natural energy of the wind and glides long distances through the air without flapping its wings. It has two key adaptations to be able to do this: firstly, special tendons lock the outstretched wings, so no energy's needed to hold them open, and secondly, the use of a dynamic flight pattern during which the bird zigzags across the waves, surfing the strong winds that are common over the Southern Ocean. By gliding at right angles to the wind, the albatross gains lift and rises to about 15 metres, whereupon it turns slightly and glides effortlessly downwind until it nears the surface of the water. Turning into the wind, the bird soars upwards once more, and the cycle repeats.

Gabriel Bousquet, a former graduate student in mechanical engineering at Massachusetts Institute of

Technology, USA, and some of his colleagues used a computer model to analyse albatross flight patterns. It turns out that by making these gentle zigzag motions as they climb and descend, drag is greatly reduced, which helps the albatross to fly more efficiently. Something else that might come as a surprise is how the bird's colouration is thought to play a role. Like many other seabirds, albatrosses are black on top, with white bellies. It was recently shown that black feathers absorb more solar energy. When they heat up from the sun, a temperature differential is created between the upper and lower surfaces of the wing. This lowers the air pressure on the upper surface, creating additional lift.

Inspired by these birds, Bousquet and his team have been working on a robotic glider that can skim along the water's surface, riding the wind like an albatross, while also surfing the waves like a sailboat. As Bousquet points out: 'If you can fly like an albatross when it's really windy, and then use your keel to sail like a boat when it's not, then this dramatically expands the kinds of region you can go.'

With that in mind, Bousquet wondered whether they could design a vehicle that could operate in both air and sea, combining the high-speed qualities and energy efficiency of both the albatross and a sailboat. The team drafted a design for a hybrid vehicle. It resembled a glider with a three-metre wingspan, which was similar in shape to the wing of an albatross. They added a tail, a triangular sail-like rudder, and a slender wing-like keel. According to their calculations, this wind-powered vehicle would only need relatively calm winds of about 6 mph (5 knots) to zip across the water at a speed of 23 mph (20 knots), which, in terms of speed and efficiency, is pretty good going.

When they built the prototype, they felt it had real potential. Bousquet's vision is that research teams could use these compact, speedy, robotic water-skimmers to survey large areas of the ocean, especially in places like the Southern Ocean, where electricity and charging batteries to power electronics comes at a premium. In an area like this, which is difficult to access, an artificial albatross would be perfect for collecting all sorts of data, from the gaseous exchange of carbon dioxide across the region to the collection of atmospheric information, both of which would help us to understand the effects of climate change and also better predict the path of hurricanes and powerful storms.

But, don't expect to see any of these features on passenger planes. Soaring like an albatross would make for a miserable flight. Even Bousquet pointed out that you wouldn't want to be on aeroplane that performed a repetitive zigzag manoeuvre, over and over, for eight hours straight. Even so, just picture it: vast fleets of albatross gliders moving up and down and in and out of the waves, silently gathering scientific information to broaden our horizons. It would be a beautiful, environmentally friendly way to survey the oceans, and the albatross gliders might well have real albatrosses keeping them company!

OWLS AND TURBULENCE

Conventional drones come in two types: those with rotors, like helicopters, and those with fixed wings, like regular aircraft. Being relatively small, the first type is easily buffeted by the wind, and drones with fixed wings are easily upset by turbulence. You're probably shuddering at the very word. Many of us have been on a plane that's experienced turbulence, as it's flown through unsteady, irregularly moving air. It can be quite a scary experience, and I've had a few eye-popping moments myself – it literally felt like I was on a rollercoaster, and that was on a commercial airliner. The smaller the aircraft – in this case the drone – the easier it is to be knocked around, and that's why experts from the Royal Veterinary College teamed up with engineers from the department of Aerospace Engineering at the University of Bristol, UK, to see whether birds might offer a solution to stabilising drones and small aircraft, so they can cope better with turbulence and sudden gusts of wind.

They chose to study birds – trained by West Country falconers Lloyd and Rose Buck – that included a hawk, a raven, a tawny eagle and a barn owl. Now a little heads up: the owl was no ordinary barn owl, oh no! Lily, as she goes by, having appeared on many BBC natural history films, was already an international television sensation, and it was her consistent flight performance that impressed the team the most.

To study these birds, the team constructed a 14-metre-long corridor. In this space, the birds would have enough room to flap their wings and build up enough speed, before settling into a smooth glide. As they glided, fans would blow out air to create strong gusts of wind that would force the birds upwards. At the same time, a clutch of ten high-speed photo and video cameras, recording at 1,000 frames a second, captured the moment the gust hit the bird. This gave our experts the ability to watch back and see exactly how Lily responded.

When Lily flew through this corridor, the team saw something that took them by surprise. Even before they analysed her flight, they could see that her body and head remained stable, despite the strong upward gusts of air. When they studied the images, it became clear why. In the split second that the gusts of air struck Lily's body, her wings immediately pivoted upwards, around her shoulder joints. It wasn't that Lily was consciously moving her wings to this position, it was more of an instant reaction, something that happened without her having to control it. This is a purely mechanical response that biologists call a preflex – a type of reflex that bypasses the nervous system altogether, and instead uses the elastic properties of a muscle to respond quickly to a stimulus. And, that's when the team came up with an ingenious idea.

They wondered if what Lily's wings were doing mirrored what happens when you hit a ball with that perfect shot; I'm thinking about sports like baseball, tennis and cricket. If you've ever played a game with some friends, or maybe you're a professional player reading this right now, I'm sure you know what I'm talking about. It's that incredible moment when the bat and ball connect in such a way that the stroke seems effortless, yet the ball bounces back with phenomenal speed... right there, you've stumbled on a magic sweet spot. In engineering, this sweet spot is called the centre of percussion.

The research team had found some truly remarkable evidence that the upward gusts of air were hitting Lily's wings at their centre of percussion, which in turn stimulated the preflex response that made Lily's wings rotate upwards, without disturbing her body. In this way, she was able to absorb the effects of the gust without being bumped around by it. This is how her head and body were able to stay so remarkably still, with all of this happening unconsciously and on-the-fly. The key point here is that this preflex gave Lily's central nervous system a few extra moments to detect what was happening, catch up and then activate her muscles in the right way, so she could maintain control of her flight.

Having made this discovery, the engineers are now exploring the different ways in which Lily's preflex response might be transferred to aircraft design. In the future, perhaps, drones and small aircraft might have some kind of hinged arrangement at the point their wings meet the fuselage, enabling the craft to respond better to gusting winds and turbulence. This could have a dramatic effect on what drones and small aircraft can deal with, making them far better at handling strong winds. And, maybe, one day, you might even see hinges like this on bigger aircraft, helping to give us a smoother, more comfortable ride; all inspired by the steady flight of Lily, the superstar barn owl.

FLY LIKE A BAT

The aftermath of an earthquake is probably one of the most devastating examples of a natural disaster we might ever encounter. Houses lie flattened, huge trees are scattered across highways, and cars have been flipped upside down as if they were play-bricks. It's complete chaos. To make things worse, there could be leaking gas pipes, live electric cables, and who knows what other dangers. This isn't a scenario where you'd want to start exploring on foot. But what if you had the ability to suddenly transform into a levitating machine that could move effortlessly and safely through the air. What might this eye in the sky look like?

You'd need to be something relatively small, flexible and agile that could fly in and out of small spaces and around tight corners – something like a bird, or better still, something even more nimble: how about a bat? Winged mammals have been the talk of the town for a team at the University of Illinois, USA, studying this group closely to help them design a flying robot, a drone essentially, but one which looks like and is inspired by a bat.

I think it's safe to say that bats haven't had the best press over the years. We associate them with witchcraft, evil, darkness, vampires, even death; but this couldn't be further from the truth, and I speak from personal experience. My first real encounter with bats was back in 2014. At the time, I was investigating the world of animal senses, including how bats navigate at night. The film crew were using infrared cameras which visualise body heat to

generate images in complete darkness. Once the cameras had been set up, I stepped into our blacked-out room and was plunged into the pitch black. The next thing I knew, ten Egyptian fruit bats began swooping around my head.

These bats are part of a family known popularly as 'megabats.' Despite this name, the Egyptian fruit bats have relatively small bodies, about 15 centimetres long, but their wingspan can be as much as 60 centimetres. They're called fruit bats because they feed almost exclusively on soft fruit. They're found throughout Africa, except in the Sahara region, and throughout the Middle East as far as Pakistan and Northern India.

It was one of the most bizarre feelings, to have my sense of vision stripped away, right when I really felt I needed my sight the most. Not only could I hear the bats, but I could also feel when they were approaching, as gusts of air from their beating wings came in from all angles. It was a little unnerving to begin with; it's not every day that you stand in a room where you can't see a single thing and you're alone with not one, not two, but ten bats, whizzing past your face. My initial instinct was to try to dodge them, but, gradually, as I became more and more relaxed, it became crystal clear these were the exact conditions these bats had evolved to thrive in.

Bats can navigate in the dark using a system called echolocation. They produce pulses of sound and listen to the echoes to create a 3D image of their surroundings. The insect-eating 'microbats' produce very high-frequency clicks from their larynxes, while many fruit bats, including the Egyptian fruit bat, produce the sound by clicking their tongues. There's even a species that produces echolocation clicks with its wings. Thanks to their ability to navigate in the dark, the bats were able to move around me in our small filming enclosure with pinpoint precision. They didn't bump into me once. Could this ability, as scientists have long wondered, offer new ways for those who are blind or partially sighted to rediscover their world?

Other than bats, mammals that are said to 'fly' – like gliding possums and flying squirrels – actually 'glide' through the air. Bats, however, belong to the order *Chiroptera*, a group of animals that have evolved a membrane of skin which covers their forelimbs and elongated fingers and toes to form webbed wings. It's this stretched piece of skin that enables the bat to sustain true, powered flight. It helps it to capture prey, avoid predators, and migrate over vast distances. And now, they're behind the design of a flying robot that could help us out in hazardous disaster zones.

If you've ever watched bats on the wing, you'll know their movement can appear to be quite erratic and graceless, but bats are, in fact, highly efficient flyers, even more so than birds. Sharon Swartz, Professor of Biology and Engineering at Brown University, USA, has suggested the secret to efficient bat flight lies in the furry creature's multi-jointed wings and flexible skin membrane, which together create a shapeshifting structure that provides more lift, less drag and greater manoeuvrability than the wings of birds. Unlike insects and birds, which have relatively rigid wings that only move in a few directions, a bat's wing has more than 20 different joints. These are covered in a thin elastic membrane that can stretch to generate lift in a variety of ways. Put simply, they're highly skilled and excellent at controlling the three-dimensional shape of their wings during flight. As Professor Swartz points out, insects can move the joint at the equivalent of our shoulder, but that's the only point they can exert force and control movement. Birds have more joints in their wings than insects, which helps, but still it's nothing compared to bats.

Bats have a skeleton that, in many ways, resembles our own, which shouldn't come as too big a surprise seeing as they're mammals just like us. What is remarkable, though, is that every joint in the human hand can also be found in a bat's wing… plus a few more. Once you know this, it becomes much easier to picture the flexibility of this wing; in fact, you can do it yourself right now. Hold out one of your hands to the side and lock your elbow into your hip. Now, slowly flex your fingers back and forth. Roll them up tight, and then expand them. Move them around at different speeds and different angles. If you think of this as the bat's wing, you can picture how they use this dexterity to make fine-scale adjustments during flight.

As for the thin and flexible wing membrane, studies revealed that the wing is mostly extended on the down stroke, when the bat is flying straight ahead, and, because the membrane can curve and stretch more than a bird's wing, bats can generate greater lift with less energy. By using a stream of non-toxic smoke and high-speed video, Professor Swartz and her team were able to see how air flows around bats as they flap their wings. During the down-stroke, the air vortex – which generates much of the lift – closely tracks the animal's wing tips, but, on the up-stroke, the vortex appears to come from another place entirely.

Seth Hutchinson, Professor and KUKA Chair for Robotics at Georgia Institute of Technology, USA, has an insightful explanation of what's going on here. 'When a bat flaps its wings,' he says, 'it's been compared to a rubber sheet. It fills up with air and deforms. And then, at the end of its down-stroke motion, the wing pushes the air out when it springs back into place. The result is that you get this big amplification of power, which comes just from the fact you are using flexible membranes inside the wing itself.'

There are also a whole bunch of muscles in the wing membranes, and bats can use these to change the stiffness of the skin in their wings, to make tiny adjustments to their flight. They also have another trick up their sleeve – or should that be 'in their wing' – for they not only bend, flex and puff out their wings, they're also able to land upside down. As they approach a tree branch, for example, they slow down, flip themselves upside down and land, hanging on to their target. It's a bit like doing a high dive, but in reverse.

Bats have such control over their wings that they can recover quickly from being buffeted by a sudden gust of wind. Professor Swartz constructed an elaborate set-up of lasers and air jets to put this theory to the test. When a bat flew through the laser beams, it triggered a puff of air, aimed directly at the bat. The team found that, even when the bats encountered strong gusts, they could recover their stability within a single wing beat. The skin of a bat is covered in sensory nerves, so it's quite likely these provide real-time information about the airflow over the animal while it's flying.

You might think all this is a tough challenge to take the bat as a model for a drone, but they've proved irresistible for the Illinois team, working also with the California Institute of Technology. The wings of their bat – known as Bat Bot – are soft and articulated in such a way to mimic the key flight movements of bats. Videos of the machine flying are extraordinary; from a distance it looks like a big white undiscovered species of bat. It's very impressive.

The robot flaps its wings just like a real bat, although, in reality, the Bat Bot is a much simpler version of Mother Nature's original design. One challenge the team faced was that bat wings use over 40 joints during flight. For a robot to have this many would've made it far too heavy and cumbersome, so instead, it has a wing with just nine joints, each made of carbon-fibre. This kept the wing light, yet strong and highly flexible.

The next challenge was the choice of material for the wing membranes. It too needed be light, strong and flexible, in order to change shape. They settled on stretchable, ultra-thin silicone-based membranes. Inside, tiny motors in its backbone controlled the robot, and it has on-board sensors that measure the angle of the joints, to help it adjust its wing position as it flies. When the team put it to the test, the robot was able to perform banking turns and steep dives, exactly the kinds of movement you see when bats are chasing prey. One of the team, Soon-Jo Chung, argues that a bat-inspired robot is a more energy-efficient and reliable solution over traditional aerial robots since stationary hovering is difficult for quadcopters – drones with four rotors – in the presence of even the mildest of winds.

What the team have designed is a flying robot that could be used to survey locations that are too dangerous for human inspection, like collapsed mines or buildings or an abandoned nuclear reactor. Whilst it's still at the prototype stage, this bat robot is not only manoeuvrable and durable, but, by supercharging it with a radiation detector, 3D camera system, and temperature and humidity sensors, the team believe the possibilities for its uses are endless. So the next time you look up and are surprised to see bats flying in broad daylight, look again. Chances are they may well be Bat Bots, surveying the landscape below.

11

Fog Harvesters

Water is essential to all of life on Earth, but while some parts of the world are fortunate enough to have water come to them – through high rainfall or just by having easy access to lakes and rivers – access to water in desert countries, in particular, is extremely limited. It's shocking to think that the United Nations forecast that 1.8 billion people will be living in countries suffering from water scarcity by 2025. I bet you're secretly thinking, 'Good thing I don't live in a desert country.' Ah, well, I've got you on that one there! Turns out, the rest of us aren't getting off lightly either, with two-thirds of the world's population forecast to be living under water-stressed conditions. This isn't a 'them' problem, it's an 'us' problem – and the effect of climate change will only serve to make things even more extreme.

So what if we could have water finally go to those to whom it didn't before? This is where the story of a little known Namibian beetle begins, one that could teach us how to extract water from right out of the air and, in the process, provide one of the major solutions needed to solve the pressing issue of global water scarcity. *Stenocara gracilipes* is the scientific name for the blueberry-sized, long-legged Namib desert beetle, a species of darkling beetle native to the Namib Desert of southwest Africa. With rainfall sparse and unpredictable, this is one of the most arid ecosystems in the world.

What this region lacks in rainfall, it makes up for in a coastal fog that drifts in with the cold Benguela Current and the early morning breeze. This ocean-powered fog can reach as far as 100 kilometres inland, and has proved invaluable to those that most know how to use it. With such little rainfall here, it's an area inhospitable to many, yet the *Stenocara* beetle wields an evolutionary trick that helps it to survive by extracting water directly from the desert fog. Our hero beetle is a 'fog-bather.'

Basking in the fog, it uses its own body to collect water. It does this by stationing itself on the ridge of a sand dune that faces into the breeze, its body angled at about 45° and its head pointing upwind. Tiny droplets of water from the fog slowly collect on the beetle's hardened wing cases or elytra. The droplets are tiny, about 15–20 micrometres in diameter; in fact, they're so small, they're invisible to the human eye. The beetle is able to collect the fog thanks to a pattern of special bumps and nodes on its back, and it's these that are vital to the entire process.

Hunter King, a physicist at the University of Akron in Ohio, USA, is someone who wanted to look at this in greater detail. Using 3D printing techniques to test several surface types in wind tunnels, he worked out which surface collected fog the best. He discovered that taking a sphere and covering it with a bumpy surface, like the surface of the beetle's back, practically turned it into a 'fog magnet.' It collected droplets of fog water at 2.5 times the efficiency of a smooth surface. Two and half times more water, just by adding a few bumps! Scale that up to the thousands or millions more litres of water that could be collected, and what a difference it would make.

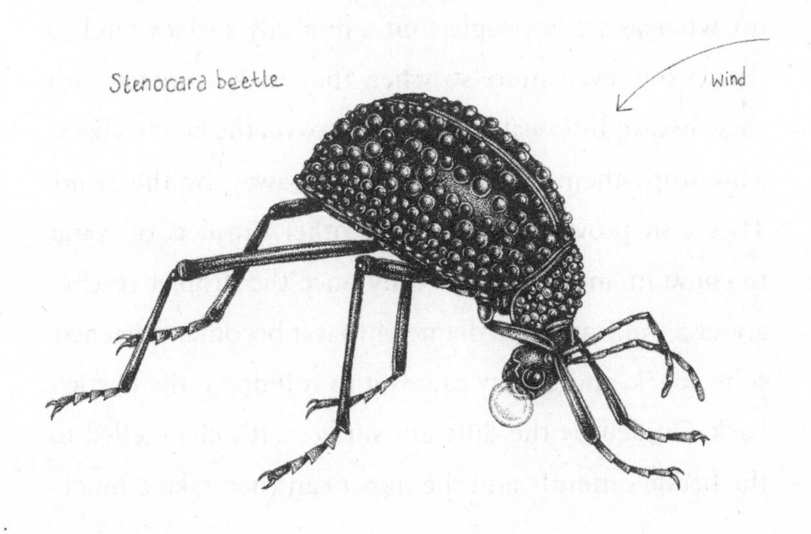

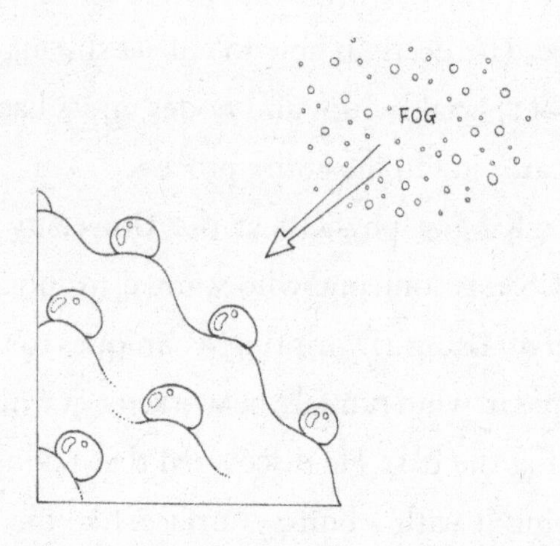

The surface of the wing cases had a few other interesting properties too. The peak area of each bump is hydrophilic, which means it attracts water molecules. The troughs in between have a waxy hydrophobic covering, so they repel water. You might not have thought about this before, but water droplets on an oily surface tend to ball up, whereas water droplets on a non-oily surface tend to flatten out, even more so when they make contact with these hydrophilic water-loving bumps on the beetle's back. This stops them from being blown away by the wind. They also provide a surface for other droplets of water to come in and attach. It's only once the droplet reaches about 5 millimetres in diameter that it becomes 'detached', so to speak, and gravity causes it to roll down the beetle's back. Guided by the different surfaces, it's channelled to the beetle's mouth, and the insect can then take a much-needed drink. This is how those clever beetles survive the harsh desert landscape: by the simple act of harvesting water from the early morning fog.

Faced with the same burden of finding reliable access to water, humans too have been collecting water from the air for some time now. The first example we'll look at dates back about 2,000 years, when the Roman naturalist and philosopher Pliny the Elder told of the inhabitants of the Canary Islands who would gather fog droplets trapped by trees with the help of collection stones they put under the branches to catch the dripping water. In Morocco, people there have also been harvesting fog using a mesh that collects water droplets that are then guided by pipes to those in need. Similarly, a story of ingenuity at its finest involved the Chilean village of Chungungo, which, between 1992 and 1997, had an impressive 94 giant mesh collectors set up on the nearby ridge of El Tofo. They really had coastal fog trapping mastered, collecting an eye-watering average of 15,000 litres of water a day, with the highest peak in a single day coming in at 100,000 litres. This water was piped 7 kilometres down the mountain to the parched community of 700 living below. Alas, the project, hailed as an environmental success story, is no longer operational, but having successfully highlighted both the potential and practicality of fog harvesting, there's no doubt its legacy lives on through projects of the future that it continues to inspire.

When it comes to harvesting fog, the most practical option is to use plastic mesh collectors. They're cheap, efficient and durable. So how to take plastic mesh, enhance it and make it fit for the future? This is where knowledge gained from carefully watching the *Stenocara* beetle may prove useful. How about designing a completely beetle-inspired new material? Might something like that be more effective at gathering fog water or dew than traditional plastic mesh?

Under an electron microscope, the beetle's tiny bumps and channels appear as huge mountain ranges and deep valleys. With this roadmap, we can now copy the microstructure and, through an injection-printing technique, reproduce it in reams of sheeting. Andrew Parker at the University of Oxford and Chris Lawrence of security and defence contractors QinetiQ, both in the UK, are both looking into this area. They see potential for these types of materials in a variety of devices. Be it water from the roofs of buildings or water channelled along a storm drain, these materials might also be useful when it comes to improving traditional methods of water collection.

Researchers at the Massachusetts Institute of Technology, USA, have replicated the beetles' ability to collect water by creating a specially textured surface, which combines the use of alternating hydrophobic and hydrophilic materials and regions. The main use is touted as being ideal for fog harvesting in places where water's scarce. Naturally the big players are eyeing up this water-collecting magic trick too. NBD Nano has shown a lot of interest in this area. In fact, you may've heard of some of their ideas: they're the people behind a concept for a self-filling water bottle, one that, depending on the environmental conditions, could collect up to three litres of water an hour. Inspired by the beetle, the surface of the bottle would be covered with textured hydrophilic and hydrophobic material. Fans would direct the surrounding foggy air to flow over the surface of an entire rack of these special bottles. Once enough fog's been collected, the captured water could be used there and then or stored away for later access.

THORNY DEVIL

The *Stenocara* beetle, of course, isn't the only creature that can teach us something about efficient water harvesting. The arid Australian outback is home to a fearsome-looking creature called the thorny devil. Despite its ominous name, this lizard eats only ants, and moves in a manner that's rather relaxed and gentle, although, with a body covered in thorny scales – a defence against would-be predators – it still looks an enchantingly formidable sight. It lives in the scrubland and desert that covers much of central Australia, which, with less than 250 millimetres of rainfall a year, is one of the harshest landscapes on the planet. Despite living in such hostile conditions, when the puddles from which it drinks dry up, the lizard somehow still finds enough water to survive.

Its secret is skin deep. Between those intimidating spikes is a subtle network of microscopic grooves. These grooves can be used to absorb water out of moist sand, drawing up the fluid, against the pull of gravity, across the

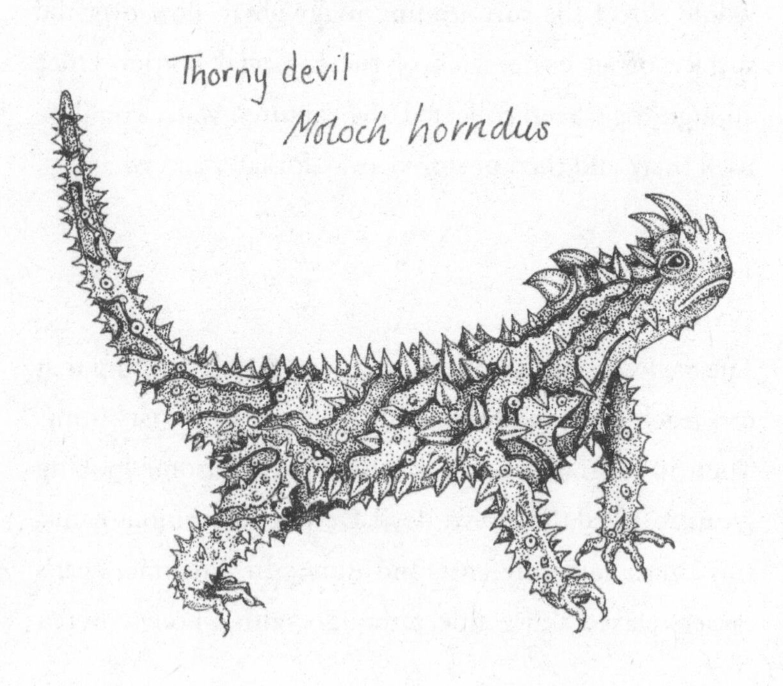

lizard's body and into its mouth. All it needs to do is stand in the right position and it drinks with its skin – the ultimate moisture harvester. And, would you believe, that's a discovery almost a hundred years old.

In 1923, biologist Patrick Alfred Buxton wrote about the thorny devil in his book *Animal Life in Deserts*. He noted how our reptile 'has the power of absorbing water through the skin, after showers of rain.' It sounds cool, although it doesn't seem to make much sense when you think about it: a desert reptile with permeable skin? When it was hot, surely that would work in the opposite direction, leaving the lizard all dried out? Then, in 1962, two scientists from the University of Western Australia discovered what's really going on. Placing one of the lizards into a shallow pool, they noticed how 'an advancing waterfront' moved over its skin, towards its mouth, which it opened and closed.

The water wasn't being absorbed directly into the skin, but was moving across the skin, through a process called capillary action. Because of the natural attraction of water molecules to the surface of a tube and to each other, water flows unassisted through very narrow tubes, as do other liquids. You see this principle at work yourself, every time you dip a sponge or a paper towel into water. The submerged parts get wet as expected, but the waterfront you see moving up and away from the water, my friends, is all down to capillary action. Acting like the paper towel, the lizard's body does the same thing. Liquid from tiny pools of water is drawn up from the moist sand through a capillary system that lies between the lizard's scales.

In 1993, Philip Withers, Professor of Zoology at the University of Western Australia, suggested the lizard was taking advantage of early morning dew condensing on the sand. It took 23 years before Professor Withers, along with Philipp Comanns at RWTH Aachen University in Germany, published a paper in 2016 that analysed and finally explained the process in more detail. When placed in a puddle, the lizards would start drinking within ten seconds, quenching their thirst within an hour. During this time, they were observed opening and closing their mouths 2,500 times, helping them to down a tremendous, wait for it... 1.28 grams of water. I know that doesn't sound like much, but in the desert every drop counts. Interestingly when they were put on damp sand, with a moisture level of 22 per cent, the thorny devils could only fill 59 per cent of the skin grooves, which wasn't enough to have water reach their mouths. To overcome this, the lizards have adapted their behaviour, opting to shovel damp sand onto their backs, a behaviour that's especially noticeable after rain showers. This increases the surface area used to absorb water, boosting the percentage in the skin grooves, and enabling the lizards to quench their thirst.

Philipp Comanns went on to study similar behaviour in Texan horned lizards, another spiky-bodied reptile found across North America, from Colorado in the USA, through to northern Mexico. Philipp compares the capillary action used to absorb water from the sand to having a straw attached to your skin. Taking advantage of the moist sand is precisely how the lizards stave off dehydration. The team have designed plastic sheets based on the skin and its capillary properties, which can slow the flow of fluids in one direction, whilst encouraging its flow in another. They seem to think there's potential for this technology to be used in electronic ink displays, food and drink production and medical appliances, to mention a few. Yet a more pressing question still stands. Is there a way to collect and distribute water in a similar way to Mother Nature, for the benefit of the billions of people who'll need it not just in the future, but need it right now? Could a thorny problem be solved by a spiky devil and a fog-basking beetle? The work in this water world continues.

12

Sharks and Hospitals

I'll never forget my first time diving with sharks. I'd made my way to the coastal community of Simon's Town in South Africa. The plan? To get as close as possible to a great white shark, the world's largest and most powerful species of hunting shark. These animals can reach up to seven metres in length, yet, despite being longer than a monster truck, it's hard to appreciate this size from above the water. Getting into the ocean with them, though, would be the perfect way of sizing them up. I was about to get a true sense of their sheer bulk, and see for myself how their stealth and speed make them such formidable predators.

Some shark enthusiasts are bold enough to dive with great whites in the open ocean, but not me. I wasn't taking any chances. I decided I'd be better off sticking with the more traditional approach that has divers stay within the safety of a cage. But, before I'd even hit the water, my mind was racing. My attention was drawn somewhere else: a video I'd seen just a few days earlier. A group diving the island of Guadalupe off the coast of Mexico somehow ended up with a diver in a cage, trapped, with a great white stuck inside. I couldn't believe the images that were unfolding before my eyes. Moments earlier, the shark, lunging for the tuna bait that was used to tempt it closer, overshot its target. As it smashed into the cage one of the metal railings broke loose, leaving the shark equally surprised at being on the wrong side of the bars. Fortunately, after what seemed like a marathon of thrashing from the shark – remember

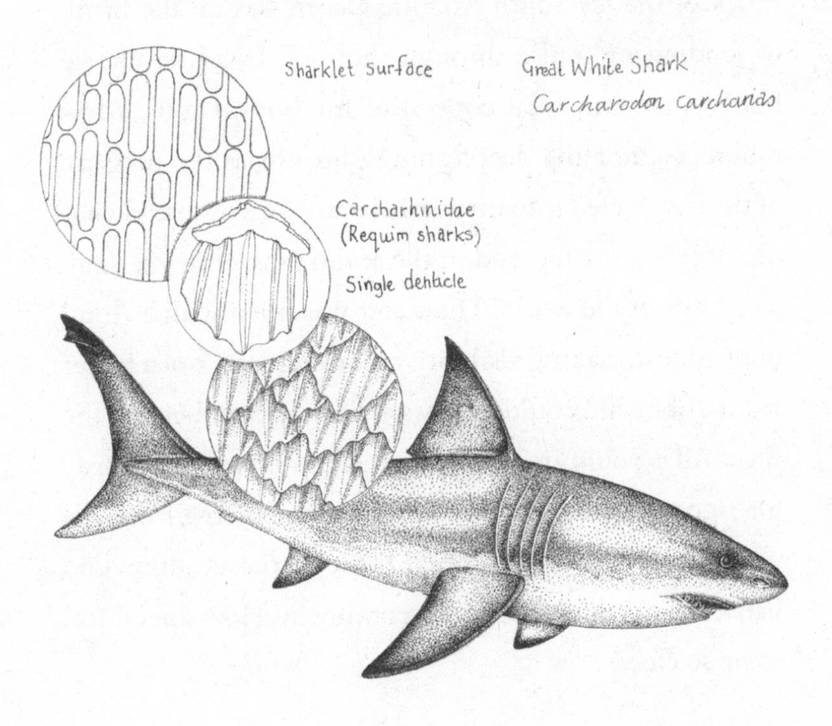

they can only swim forwards – our great white eventually escaped out the top of the cage, with only a few minor injuries. And what of the diver inside? To my amazement – and yours if you get round to watching that video – he was totally fine, physically shaken in the cage, but not in any way stirred by the experience. In fact, he seemed inexplicably unfazed. Still, you can understand why I – now confronted with my own shark experience – had my heart well and truly in my mouth.

With all our checks complete, I entered my new caged office. As I plunged beneath the surface of the water, the shock of the icy South Atlantic Ocean was on the brink of sending my pulse through the roof. I had no choice but to maintain total control of my body. I took a few moments, inhaling deep calming breaths as the pressure of the oxygen fed into my mask. I was now completely out of my comfort zone, and in the realm of the biggest predatory fish in the world. The water was filled with a cloud of plankton, making visibility seriously, for lack of a better word, rubbish! I could barely see a few metres beyond the cage. All I could do was float there and wait cautiously for signs of this 'stealth of the deep', to emerge from the murky blue. A voice crackled through the headset: 'OK, Patrick! Get ready, the shark's coming in! How does it feel to be so close?'

I replied with silence. I still couldn't see anything. Wondering if they were playing an ill-timed prank, it hit me, quite literally. Although I couldn't see the shark, I could feel something big banging against the cage. I turned around and there it was: a great white shark, pushing up against the metal bars, lifting my cage up into the churning water above. There was no mistaking its power. With every flick of its tail, the cage was dashed into the side of the boat, over and over again. This was not the time to be overwhelmed by the situation, though; I had a job to do. I composed myself. With that, knowledge about this apex predator began pouring out to the underwater camera; how gigantothermy helps it maintain body heat, its preference for hunting at dawn and dusk, fur seals being its favourite snack. It was pure adrenalin. After what felt like an eternity, yet also a split second, the churning of the water eased, the shark slowly disappeared, and with that, a breath of relief.

The great white is just one of 500 or more species of sharks that inhabit oceans and some of the planet's major rivers too. They range in size a fair bit, going from the whale shark – the world's largest living fish, which can reach a length of 12 metres – to the deep-sea dwarf lantern shark that comes in no bigger than the size of your hand. The diversity of sharks we see today all evolved from a common ancestor dating back to over 400 million years ago. We're talking of a time long before dinosaurs roamed the Earth. Yet, while fossils are all that remain of these giant reptiles, sharks have survived at least five major mass extinction events to be found in waters throughout the world, still to this day. Great white sharks are truly patrons of a prehistoric world.

While the whale shark and the basking shark (the runner-up in size) are both harmless filter feeders, many other large sharks are, indeed, apex predators, right at the top of their food chain. Even then, their safety isn't guaranteed: some of these predators succumb to being preyed upon themselves. Orcas are the only ones known to challenge and take out great white sharks in such a way that just the presence of a pod alone has been known to change the behaviour of the sharks, causing them to leave an area with orcas entirely. Killer whales aside, the hunting prowess of a great white shark is second to none; after all, their bodies have been fine-tuned over millions of years of evolution and, just to rub it in for the rest of us, they've even got a sixth sense – electromagnetism. It's an epic sense to have. They use it to feel their way around the Earth's magnetic field, and also in the moments just before a strike, when they detect the minute electrical currents in the muscles of their prey. Just as well really, for in the final moments of an attack, the sharks are often swimming blind. To protect their eyes from injury, they'll either roll their eyes back into their sockets, as we see in great whites, or they'll use special membranes that come across the eyes, as we see in tiger and bull sharks. These are a group of marine animals that are especially noted for having an acute sense of smell. All this is combined into a sleek, torpedo-shaped body that makes many sharks highly skilled hunters.

Sharks also appear to slide effortlessly through the water, and for some species moving isn't really much of an issue, undertaking vast migrations to breed and find food. Female great whites, like the one I encountered in South Africa, have been tracked travelling across the Indian Ocean, all the way to Western Australia and back again. What I want to know is what it must feel like to navigate in that much space. Added to that, some sharks migrate great distances between deep and shallow water, and they do it every day. These are known as vertical migrations and, depending on the species, they can range hundreds of metres below the surface. Blue sharks, for example, spend the night near the surface of the open ocean, but then, during the day chase their vertically migrating fish and squid prey, diving to depths over 400 metres. So, it's decided: sharks are cool, with some killer qualities to boot!

And there's one more adaptation we seem to be learning a lot from, but it's a trait that isn't so obvious.

Back in my diving cage, something I noticed about the skin of the great white was how spotless it appeared: no sign of algae, barnacles – nothing. It looked so silky smooth you could just touch it. Needless to say, I did not try to stroke the shark. I kept my hands well inside the cage, thank you, but, if I had stretched out my hand to touch it, I'd have had my senses greatly aroused. If you imagine stroking a great white from head to its tail, what you'd feel would be smooth skin, but stroke the other way, and you're greeted with the jarring sensation of rough sandpaper. This is because a shark's skin is made up of millions of tiny V-shaped scales, called 'dermal denticles'. You can think of these as micro-teeth, flexible layers of small triangles, marked with grooves running down their length in line with the flow of the water over the shark's body. The grooves are there to stop swirls of turbulent water – called eddies – from forming. They would act like a brake, and slow down the shark. The grooves have the effect of letting the water flow more smoothly over the skin. It's this reduced friction that helps to increase swimming efficiency. It'll come as no surprise to learn, then, that this has made a big splash with our most competitive human swimmers.

OK, for this one, we need to go back to turn of the millennium. It's the year 2000, and international swimwear company Speedo has recently introduced a range that imitates the effect of shark denticles. The simulated shark denticles were so effective that, at the 2008 Beijing Olympics only eight years later, swimmers who were wearing a fine-tuned version of this sharkskin technology won 98 per cent of the medals. These types of swimsuit have since been banned from use in Olympic competition. Has a rebellious ring to it, right? They were so good, they had to be stopped cold in their tracks, but added speed and efficiency through the water wasn't the only benefit.

Other than scars from mating encounters and fights for dominance, sharks generally have a flawless skin. Looking at other large marine animals, many of them are host to all manner of hitchhikers, from barnacles to microscopic algae, and it's not just animals that are affected. In 2002, the US Navy had an issue of grime on their hands: their ships, especially their submarines, were coated with thick algae. Getting rid of the gloop was a matter of national security. Something had to be done.

Professor Anthony Brennan at the University of Florida is a specialist in materials science and engineering. He also happens to be part of the naval research programme which dealt with biofouling events like this. He and the rest of the team were chasing down a solution to the algae problem. He also had the challenge of reducing dependence on toxic anti-fouling paints, making vessels more efficient in the water, and slashing the number of expensive vessel dry dockings. Quite a challenge.

The first step was to understand how fouling of ships started. Bacteria, which naturally cling to the surface of a ship's hull while in search of food, excrete a variety of substances in the process. Other organisms become attracted to the area and start to attach themselves. The bacteria replicate, and the population of microorganisms starts to increase rapidly. Other organisms, including barnacles and algae, then take up residence. As they too begin feeding, they also start to grow in number. It's the equivalent of transforming a small fishing village into a towering coastal megacity. The more it grows, the more space there is for other organisms to attach, and the bigger becomes the area affected by biofouling. If left unchecked, this coating of marine microorganisms would lead to significant drop in efficiency, and the costs of combating that are pretty hefty.

Anthony Brennan, amongst other colleagues, watched as an algae-coated submarine returned to port. He made a note of how it looked like a 'whale, lumbering into the harbour.' This got him thinking: what about marine animals? Are there any slow-moving inhabits of the ocean that are also resistant to the build-up of micro-organisms? Well, you've probably guessed it by now: the answer is yes. The shark! Brennan wanted to understand what it was about sharks that stopped this growth from happening on them. Taking an impression of a shark's skin scales – the dermal denticles – and analysing this using electron microscopy, he was able to reveal the distinct configuration in which the scales were arranged: a clearly defined diamond pattern. At the magnification afforded by the electron microscope, each scale appears lined with millions of tiny ridges. Brennan had never seen anything like it. He was sure the presence of these dermal denticles, unique to sharks, was the key to inhibiting bacterial and algal growth – he'd found his pollution solution.

The dermal denticles work by preventing micro-organisms from settling on the skin. Bacteria, in this example, find it tricky to attach to this type of surface. They either have to straddle the ridges or bend into the gaps. All of this strain puts tension on their cell membranes and, compared to a flat surface, it reduces the area of contact. The energy requirements to cling on to these uneven surfaces are too great, and so the bacteria settle somewhere else. Brennan measured the width to height ratios of the ridges, to find out the exact numbers that would discourage microorganisms from settling. From this, a synthetic surface was designed, based on his findings. They called it 'Sharklet.' Armed with this new shark tech, Brennan and his founding company aim to develop a range of products that stop bacteria sticking not only to marine vessels, but on other surfaces too.

A groundbreaking application for this kind of bug-proof surface would see them rolled out in our hospitals, where keeping surfaces free of bacteria is a real issue. It's something that scientists have already investigated: how to use different materials to prevent the spread of bacteria. Copper alloys, for example, are one option – being toxic to bacterial cells. They work by interfering with the cellular processes, killing them if in direct contact. The material modelled on sharkskin works a bit differently: instead of killing bacteria, it simply stops them from attaching in the first place. You can see how useful something like this would be for any high-contact surfaces in public areas: door handles and panels, light switches, and desktop surfaces. These are everyday contact points touched by many people over the course of a day. They're great for bacteria, especially if combined with tardy cleaning schedules, but bad for us. By upgrading these surfaces with bacteria-resistant materials, we'd have the potential to reduce the spread and the infection rates of killer superbugs, such as methicillin-resistant *Staphylococcus aureus*, known to you and me as MRSA.

Back out at sea, where this story began, sharkskin technology could end up on the hulls of ships, underwater research vessels and deep-sea robots. It would finally see their surfaces free of those slackening barnacle hitchhikers. Although we often think of sharks as killing machines, with sharkskin technology for hospitals likely to be used in the not-too-distant future, *Jaws* could be the key to saving millions of lives.

13

Explosive Back End: Bombardier Beetle

The great British naturalist Charles Darwin is considered by many to be a founding father of the study of evolution. He was curious about many things: the origin of life, animal behaviour, and also how animals tasted. During his time as a student at the University of Cambridge, Darwin was a member of a cubbyhole society called the Glutton Club. This club had a niche, yet focused, agenda: to meet every week and feast on 'strange flesh.' The students tried everything from hawk (not too bad) to brown owl (downright disgusting). If you're feeling a bit shocked by what you're reading, join the queue; this is not the picture we're normally painted of Darwin's most well-known exploits. Nevertheless, his taste for the odd did continue through his adult years and into the time of his voyages aboard HMS

Beagle. In fact, he tried much on the exotic animal menu that, today, would leave many a naturalist gobsmacked – armadillo, puma and even giant tortoise, which apparently had a buttery taste. With at least three of its species having already been driven to extinction, today the giant tortoises of the Galápagos are protected, and eating them won't be winning you any brownie points.

Darwin was also a great a collector of beetles, and I have no doubt that he tasted a fair handful in his time. Eventually he got a taste of his own medicine, as he wrote in a letter to a dear friend of his. It described his experience of attempting to hold a single bombardier beetle between his teeth, as he reached to collect another specimen out in the field. Who knows what he expected the poor bug to do. What he certainly didn't expect was for it to send a stream of 'acid' firing into the back of his mouth. The 'little inconsiderate beast', he wrote. After hearing that story, I've decided that the look on a friend's face as they spit out an unsavoury mouthful shall henceforth be called 'Doing a Darwin!' Little did our esteemed naturalist know that, many years later, this fiery mouthwash would come in very handy for the modern-day technology of fuel injection and its systems of the future.

Named after the soldiers who operated such artillery, the bombardier beetle is a miniature living cannon, a mobile gun that blasts out clouds of acrid liquid, to stave off the beetle's attackers. It sends ants, frogs, spiders and even birds ducking for cover. But how is this happening in such a small space, and why isn't the bombardier left charred to a crisp?

Scientists want to study the exact conditions of what happens inside the beetle, and understand what helps it produce such an explosive jet without blowing up its own rear end.

There are more than 500 species of bombardier beetle in the world. Whilst rare in Europe, the beetles are common in Africa, Asia and the warmer parts of North America. They've been found in a variety of habitats, including forests, grassland plains and deserts. Many of them, both as adults and as larvae, are carnivorous, and typically hunt down other insects at night. They lay their eggs underground in decaying plant material or in the remains of dead animals. The newly hatched beetles go through several stages of moulting before they finally reach maturity. Some larvae even parasitise the larvae of other beetles.

The most striking thing about these armoured insects is their chemical defence. Many beetle species use nasty chemicals to defend themselves, but the bombardiers take this chemical warfare to a whole new level. Some emit their discharge as one continuous stream, mist or froth, whilst others fire it as a high-velocity pulsed jet, a boiling spray that can reach an incredible 100°C, the same temperature as boiling water; so, how does it all work?

Inside the back end of the beetle, near the tip of the abdomen, are two defensive glands, and each gland has two linked chambers: an inner chamber acts as the reservoir, with the outer chamber being used for the reaction. It's this arrangement that's been the focus of some interesting research. Teams from the Massachusetts Institute of Technology, the University of Arizona and the Brookhaven National Laboratory in the USA used high-speed X-ray imaging to observe what's happening inside the beetle.

They found that the inner reservoir, which carries the reactive chemical agents – hydrogen peroxide and hydroquinone – is separated from the combustion chamber by an inlet valve. When the beetle senses a threat, the muscular walls of the reservoir chamber produce the chemicals. At the same time, they also open the valve and the active ingredients flow into the reaction chamber, along with an enzyme that works as a catalyst. The enzyme causes the two chemicals to react in a violent explosion, producing oxygen, water vapour, irritants called quinones and, as you might imagine, a lot of heat. This noxious, steaming cocktail forces its way out through an exit valve as a spray, which the beetle aims towards those intruders. This toxic steam can be blasted up to 20 centimetres – not bad for a beetle that's less than 2 centimetres long.

So – I bet you're still wondering – how come the beetle isn't cooked alive in the process? It turns out, these beetles have evolved a way to avoid this: the reaction chamber is lined with the insect equivalent of a 'blast wall', which makes it much easier for the vapours to surge through the desired exit, rather than leaking into its body and risking a beetle explosion.

X-ray images taken at Argonne National Laboratory in the USA revealed that the opening and closing of the inlet valve between the chamber holding the chemicals and the explosion chamber seem to take place passively. It's the increase in pressure during the explosion itself that expands the membrane, and this opens the second exhaust valve, while closing the inlet valve. Once the liquid's ejected and the pressure reduces, the membrane relaxes back to its original position. The exhaust valve closes again, and the inlet valve opens once more, allowing for the next build-up of pressure. It's the chemical reaction that's responsible for heating the liquid to boiling point. This pushes the pressure up again, and opens the exhaust valves once more. The setup means reactions can happen in rapid succession – a staggering 400–500 times a second. Now that's super-fast.

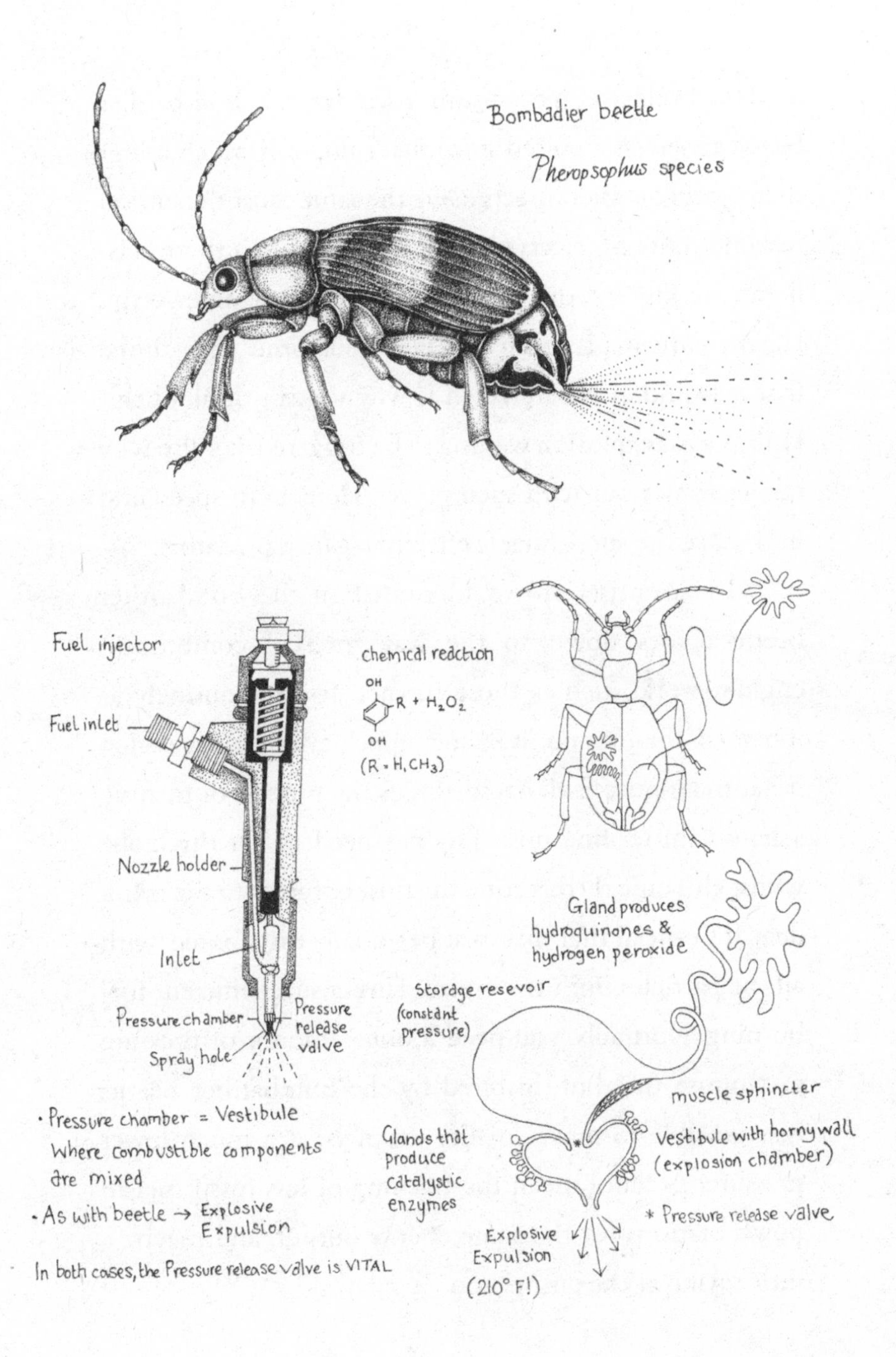
Bombadier beetle
Pheropsophus species
Fuel injector
Fuel inlet
Nozzle holder
Inlet
Pressure chamber
Spray hole
Pressure release valve
chemical reaction
OH
R + H_2O_2
OH
(R = H, CH_3)
Gland produces hydroquinones & hydrogen peroxide
Storage resevoir (constant pressure)
muscle sphincter
Glands that produce catalystic enzymes
Vestibule with horny wall (explosion chamber)
* Pressure release valve
Explosive Expulsion (210° F!)
• Pressure chamber = Vestibule where combustible components are mixed
• As with beetle → Explosive Expulsion
In both cases, the Pressure release valve is VITAL

The explosive mechanism used by the bombardier beetle generates a pulsed spray that's not only much hotter than sprays of other insects using the same caustic chemical reagents, but also ejects scalding steam. From a thermodynamic standpoint, that last part is particularly noteworthy. For the same mass, steam occupies 1,600 times the volume that it would as water, which is why we can think of this H_2O as a gas explosion waiting to be triggered. It's also why the jet comes out with such power. Here, both speed and heat make the spray hugely effective against predators.

Why scientists are so interested in the bombardier beetle comes down to the way internal combustion engines work, such as those in cars. To run routinely as intended, there's a point in an engine's cycle during which it has to atomise fuel. Atomising is the process of turning a liquid into a fine mist. Engines need to get this done whilst also quickly directing the mist of fuel into the cylinders. It's crucial that this mist be equally distributed, with all the particles uniform in size. This ensures efficient fuel burning. Normally, you need a huge amount of pressure to atomise fuel but, inspired by the bombardier beetle, is it possible to make smaller droplets, at a much lower pressure? It could mean the burning of less fossil fuel in power stations for the same energy output, and maybe a little saving at the pumps too.

The system used by the beetle is called 'pulse combustion.' Human engineers had created a variety of pulse combustion engines before we even knew what was going on inside the bombardier beetle. The infamous V-1 flying bomb, also known as the doodlebug, used a kind of pulsed combustion engine. What's unique about the beetle's system, though, is if it's accurately mimicked by scientists, we could produce a spray where we have full control the droplet size. While the beetle relies on a passive system, a team of scientists led by Andy McIntosh at the University of Leeds, in association with life science company Swedish Biomimetics 3000, has introduced an active system.

Traditionally, fuel-injection technology atomises fuel by forcing it through a sieve at very high pressure, but producing such pressures consumes a lot of energy. Technology based on the bombardier beetle, on the other hand, would enable the fuel to be atomised at much lower pressure, which means less energy to produce.

Car engines are also damaging to the environment because of the emissions they produce. A fuel-injection system based on the bombardier beetle's system would produce smaller droplets of fuel than those we can make at the moment. Because of their greater surface area, these smaller micro-droplets would burn more efficiently, giving back improved engine performance, lower fuel consumption and reduce greenhouse gas emissions. Although many countries will, undoubtedly, make the move to electric vehicles within the next decade, I think there's still huge potential for the improvement of internal combustion engines thanks to the bombardiers.

Work in this area undergoes continued development with the University of Leeds team now looking into how the bombardier beetle could inspire fire extinguishers capable of streaming out retardants to a much greater distance, potentially as far as 30 metres. This would benefit firefighters and other emergency users massively, because they wouldn't have to get so close to a fire they were trying to put out. A practical super-soaker, and maybe one that we might need soon to combat those wildfires we're witnessing more and more frequently.

We've certainly come a long way from Darwin's taste for exotic cuisine, but, it goes to show how a bad taste can turn out to be good news for the world of science... Oh, and a word of warning: if you ever come across a bombardier beetle, do yourself a favour, and don't 'Do a Darwin!'

14

Wind Farm Animals

My very first encounter with humpback whales is enshrined in my memory as one of those unforgettable moments of a lifetime. I was off the western coast of Australia, back in 2017, and I'd never seen so many whales in one place before. I watched from the boat as whale after whale passed by, greeting us with a puff of saltwater mist from their blowhole as they surfaced briefly for air. The most spectacular sight, though, was seeing them do a deep dive: rising to the surface, they'd arch their backs, and then descend into the waves, their huge tail flukes held aloft, before slowly submerging out of sight. The name 'humpback,' as I'm sure you've always wanted to know, comes from how this arching movement accentuates the hump in front of the dorsal fin.

Our location was Ningaloo Reef, an ecosystem teeming with life that draws its name from the aboriginal phrase for deep-water. I like to think of it as the Great Barrier Reef's younger brother or sister and, just like its older sibling when it comes to biodiversity, the waters of Ningaloo are amongst the richest on the planet. It also serves as a migration route, a superhighway for humpbacks travelling up from Antarctica to the warmer waters of Southeast Asia, and, like trucks on a speeding highway, these beasts are fully fuelled and rarely stop for anything.

So there I was cleaning my dive gear at the back of the boat when, out of nowhere, there was the loud, gushing sound of misty air rushing from a blowhole. The crew fell silent, and we all looked over to one side, where we were greeted by three humpback whales, just lying there, horizontally on the surface of the water. We snapped back to reality. Everyone sprang into action. Chaos and excitement filled the air, adrenalin pumping through our veins. Even the captain of the boat looked shocked, and that's when I realised we were witnessing something very special.

The crew had never seen this kind of behaviour up close before. Lying on the surface like this is called 'logging,' because the whales look like huge logs floating in the water. It's not entirely understood why the whales do this, especially in the middle of a migration route, but it could be that the whales were simply entering a state of a rest and taking a quick pit stop, which, when you think about it, would make total sense: a quick nap to recharge before getting back on the road for the long journey ahead.

The three whales stayed with us for about ten minutes, just enough time for to squeeze into a wetsuit and slip into the water. What struck me was the size of the pectoral flippers on either side of their body. They were massive – about a third of their total body length – and were lined with knobbly edges. To me, this knobbly edge seemed counterproductive to moving through water swiftly. You'd expect them to have flippers that were smooth and streamlined, but these bumpy flippers worked really well. In fact, we might end up learning a thing or two from them in designing new-age wind turbines. With the increasing demand for energy, and a burgeoning need to utilise environmentally friendly renewables, it's a whale of a thought to think humpbacks are the ones to help us out.

Despite their huge bulk, humpback whales aren't the biggest whales in our oceans. With a maximum length of 30 metres, that title goes to the blue whale, the largest animal ever to have lived. Still, humpbacks are no shrinking violets and can easily reach the length of a city bus, with some growing to 16 metres and weighing 36 tonnes. Yet, despite their enormous size, they are truly graceful in the water, where they have the ability to make sharp, tight turns, and can speed along at 15 mph, if they're in a hurry. They might be the size of a bus, but they move like ballerinas.

The story goes that Professor Frank Fish, at West Chester University, Pennsylvania, was out shopping for a gift, when a sculpture of a humpback whale caught his eye. He pointed out that the sculptor had put the bumps on the wrong side of the flipper. The shop manager quickly corrected Frank. She was familiar with the sculptor's work and knew he wouldn't have made a detailed error like that. Frank was taken aback. If the artist was right, then surely the science of fluid dynamics must be wrong. Anyone who's studied aerofoils (like wings) and hydrofoils (like flippers) will tell you that leading edges need to be smooth and streamlined. In which case, why do humpbacks have bumps on the front of their flippers?

After years of research, which started with a washed-up whale on a beach in New Jersey, Frank Fish has uncovered their secret. The bumps help water to flow over the flippers more smoothly, which gives this giant mammal the ability to swim in tight circles, and this is important because of the agility needed, to catch their prey.

Humpbacks are baleen whales so, instead of teeth, they have hundreds of bristle-like structures lining the upper part of their mouth. These are the baleen plates. As they swim towards their prey, they take in an enormous mouthful of water, along with the food. With the help of the baleen plates, they're able to easily sieve out the shrimps, fish and krill that they eat from the seawater. What's also captivating is just how agile they are when feeding, and how they use those extremely mobile flippers for banking and turning. Scientists have observed how they use their pectoral flippers as biological hydroplanes, in what's referred to as the 'inside loop' behaviour. The whale swims away rapidly from a shoal of fish with its flippers extended, then rolls 180 degrees, making a sharp U-turn before a last-minute lunge, swallowing hundreds of unwary prey. As an alternative, humpbacks will dive deep, and then swim up in a spiral pattern releasing a steady stream of bubbles from their blowholes as they go, in a process called 'bubble netting.' This is probably my favourite hunting technique: as the bubbles rise, they form a type of net, a wall of bubbles that surrounds their prey – in this case, a massive shoal of fish. The whale then pivots with its flippers, turns and swims up through the bubble net to engulf any fish trapped inside. The precision with which humpbacks accomplish these manoeuvres is down to those bumps, or, as they're more accurately known, tubercles.

To understand how the whale uses its flippers when it swims, imagine you're in a moving car on warm windy day. You stick your hand out of the window and start tilting it at different angles; when you angle your hand upwards, the wind pushes your hand upwards too. That's what we call lift – it's the same lift that allows aeroplanes to take off or the blade of a windmill to turn. Tilt your hand too far, though, and eventually it stops going upwards. This is what's referred to as a stall. Your hand now has no lift and, even worse, it's being pushed back on from the resulting aerodynamic drag. The whale's flipper acts in the same way. Like a wing on an aeroplane, it produces lift by angling the flipper into the water flow at a chosen angle of attack. In the case of the whale, the lift force is directed to help the whale turn, like a banking aircraft. If the angle of attack is too high, a wing would lose lift and the aircraft stall, but the whale's tubercles channel water in such a way that they work to prevent a stall, which would prevent the whale from making such a tight turn.

Professor Fish and his team engineered artificial flippers, equipping some with tubercles and leaving others without. The models were tested using a wind tunnel, and, sure enough, Frank's experiments showed the tubercles provided flippers with more lift and less drag, while allowing the angle of attack to be increased by 40 per cent before the flipper 'stalled'.

The small bumps cause stalling events to happen at different points along the flipper instead of right across

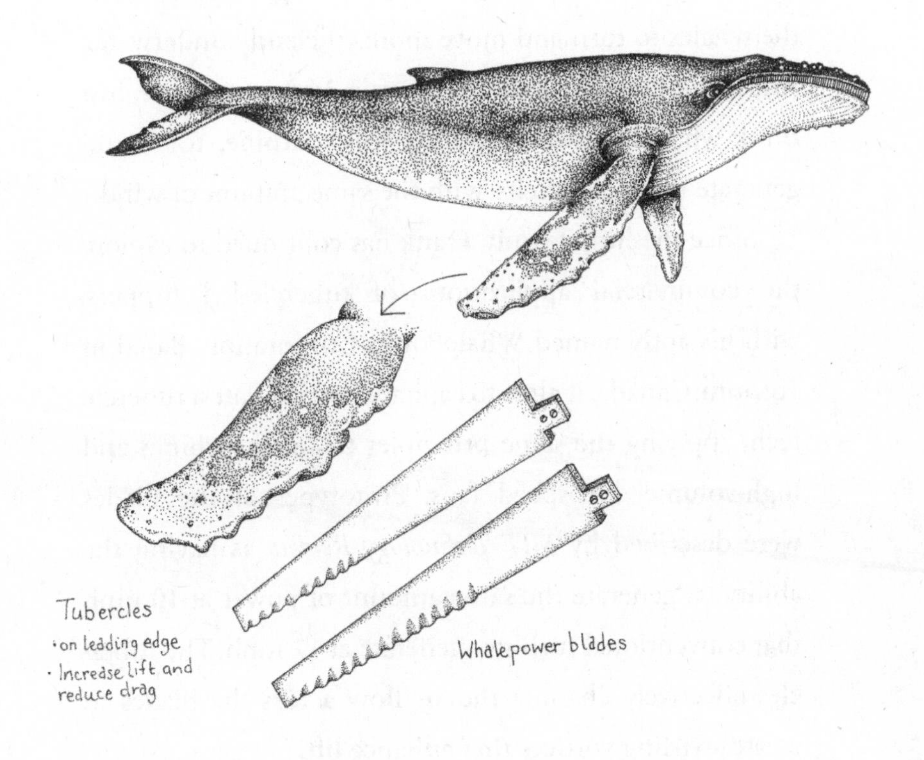
Humpback whale
Megaptera novaeangliae
Tubercles
· on leading edge
· Increase lift and reduce drag
Whalepower blades

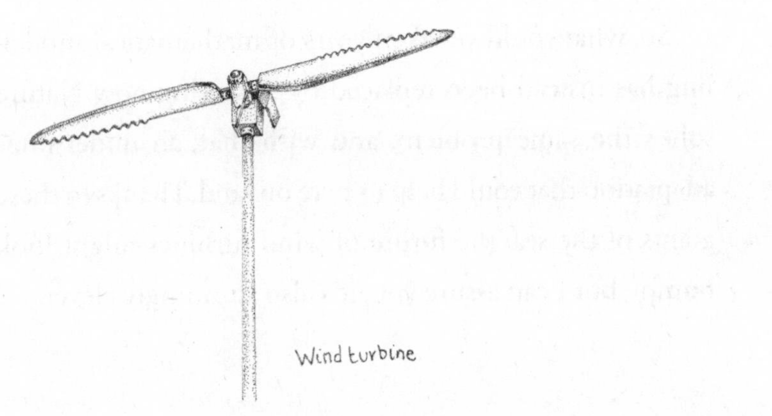
Wind turbine

it. This makes a full-on stall much easier to avoid. By increasing maximum lift, and decreasing drag, it allows the whales to turn and move more efficiently underwater. And if flippers like this can move a 36 tonne whale, just think what they could do on a wind turbine. You could generate more electricity with the same amount of wind.

Since the initial study, Frank has continued to explore the commercial applications of tubercle-led flippers, with his aptly named WhalePower Corporation. Based in Toronto, Canada, it aims to capitalise on this latest tubercle tech, applying the same principles to small turbines and high-volume, low-speed fans. Prototypes of the blades were described by *MIT Technology Review* as having the ability to 'generate the same amount of power at 10 mph that conventional turbines generate at 17 mph. The tubercles effectively channel the air flow across the blades to create swirling vortices that enhance lift.'

So, what could've taken years of mathematical modelling has instead been replaced by observing how Nature solves the same problem, and with that, an underwater adaptation that could help us here on land. Thanks to these giants of the sea, the future of wind turbines might look bumpy, but I can assure you, it's also stunningly clever.

BACK TO SCHOOL

Some of the sea's smaller creatures are also contributing to the harnessing of wind power, and this time we have our detective lens fixed on fish – fish which spend most of their lives living in big shoals and shimmering schools. One of the most perplexing yet striking underwater sights has got to be a school of fish swimming and flashing in unison. In some cases, there are hundreds, even thousands, of individuals, but packed together like this, they all appear to move as one. Twisting and turning, the group contracts and expands, dividing and reuniting – fission and fusion. But, how does any of this relate to the design of wind farms?

There are two sorts of fish congregations: shoals and schools. A shoal describes a group of fish that hangs out together but aren't organised. A school of fish, on the other hand, is highly structured, with coordinated movements, which results in all the fish travelling in the same direction at the same time. Just to confuse things, a group of fish can switch from shoaling to schooling and back again.

Schooling, specifically, seems to have evolved as a defence mechanism to being preyed upon. What's surprising about a school of fish is that there's no leader. Instead, each fish follows two simple rules: stay close to your immediate neighbours, but not too close, and keep swimming. When a single fish turns, its immediate neighbours turn, and then their immediate neighbours turn, and so on – think of a Mexican wave, but for fish, and in 3D. The synchronised fish move left and right, forwards and backwards, and up and down. This has the effect of making the school appear as if it's a single moving organism. So, how are all these fish organised?

Every fish has what's called 'a zone of repulsion,' which might sound like a place you go when you get friend zoned, but in fact describes how fish automatically turn away from their neighbours to avoid a collision. Outside of this zone is the 'zone of orientation,' where each fish attempts to match its neighbour's movement. When the school's moving together, each fish needs to orient itself such that it matches what its neighbours are doing. When the school's stationary, maintaining a close distance between individuals becomes an even bigger priority. There's safety in numbers, and this is most apparent in a natural underwater phenomenon known as a bait ball.

I first came across these bait balls while watching a wildlife documentary called *South Pacific*. I'd never seen anything quite like it before, and I was blown away. Like a huge underwater tornado, a great turning column of shimmering fish was swirling in the sea, thousands of them, all swimming for their lives. Literally. Their aim is to confuse any approaching predator with quick flashing movements. The chances of being caught in such vast numbers are much less if you're one of a crowd. In some cases, however, this avoidance technique can end up working against the fish. Some predators, such as sea lions, work together to divide the fish into smaller bait balls. Isolated from the main group, the sea lions take turns to swim through the ball and feast on the helpless fish.

Schooling is also a great a way to save energy. If a fish is swimming by itself, the energy it uses to push through the water is dissipated into the surrounding water, but if another fish follows closely behind, it can utilise the energy the leader has already expended by moving into its wake, which makes it easier for the fish behind to move forward. It's a bit like a slipstream. As a result, fish at the back of a school, not only beat their tails at a lower rate, but they also use less oxygen compared to a lone fish.

Daniel Weihs, at Technion Israel Institute of Technology, suggested that the most efficient pattern for a school of fish – a flat diamond arrangement – takes advantage of the vortices produced by the two fish that are diagonally in front of any given individual. In this diamond pattern, it's theoretically possible to save up to an impressive 80 per cent of the energy needed to swim. Surprising as it may seem, scientists have been studying this water-based behaviour, to help the air-bound behaviours of wind farms and ultimately the farms' design.

Most modern wind farms are arranged in rows, with propellers above spinning on gleaning white vertical poles. You've probably seen one from a distance, but it's only once you're up close that you realise just how big they are. Viewed from directly below they seem to reach like outstretched arms to the heavens. And, it's because of their size that wind turbines have to be spaced a fair distance from one another, to stop the turbulence from one turbine adversely affecting the efficiency of another. This is partially solved by using bigger blades and taller towers, but then you run into the problem where you start generating higher noise levels, and you increase the chances of the taller tower becoming a hazard to birds and bats.

At the Californian Institute of Technology in Pasadena – aka Caltech – researchers decided to approach the problem in a new way. They looked at the design of the wind farm itself, focusing on what was happening nearer the ground. The challenge, as they saw it, was to maximise energy-collecting efficiency closer to the ground – at around nine metres, instead of the current target height of 30 metres. As the engineers explained it, the global wind power at nine metres is several times more than the world's electricity usage, so, in theory, with the right design, smaller and less intrusive turbines arranged in the right way, should generate plenty of energy.

In the Californian desert, John Dabiri, now at Stanford University, led a team that looked into building an array of new wind turbines, only these weren't oriented horizontally, like regular propeller-styled turbines, but were set up on a vertical axis. John and his windfarm gang soon found their field lab became home to two dozen '10-metre-tall, 1.2-metre-wide vertical-axis wind turbines'. With their sweeping vertical rotors, they've been described as looking like giant egg whisks sprouting from the ground. What's really special about them, though, is their positioning and relationship to one another. The pattern is based on the swimming behaviour – and fluid dynamics, as it's called – of schools of fish.

If these turbines are positioned very close to one another, they actually gain a greater opportunity to capture all of the energy of the prevailing wind, and even wind energy from above the farm. Having each turbine turn in the opposite direction to its neighbours also increases their efficiency. The opposing spins help air to move more freely through the windfarm, reducing the drag effect on each turbine, which means they're free to spin faster. Each turbine funnels air to its neighbour without a loss to turbulence.

In the summer of 2010, John Dabiri carried out some field tests, measuring the rotational speed and power generated by six turbines when placed in different configurations. The tests revealed that having the turbines arranged so they're four turbine diameters apart (roughly five metres) completely eliminated aerodynamic interference between neighbouring turbines. In order to achieve the same thing with propeller-style turbines, they'd need to be spaced about 20 turbine diameters apart. Based on the largest wind turbines now in use, that'd be equal to a staggering spacing of 1,600 metres between towers. So, vertical-axis-wind-turbines seem to have several advantages. They can be placed much closer to one another than traditional wind turbines, and they capture wind energy from multiple directions, even from wind movement above.

Caltech reported that these six 'counter-rotating vertical axis wind turbines' generated 21–47 watts of power per square metre, which claims to be ten times that of conventional wind farms, plus, with turbulence reduced, the turbines at the back of the farm array were still able to generate 95 per cent of the power of those in the front row. There's no doubt this is interesting work, and, with still so much to learn, could it be that the moves of schooling fish will be the decider on whether 'vertical' wind farms will be the wind farms of the future?

15

Hedgehogs and Helmets

It's hard not to love a hedgehog. With their sharp spines or quills, their soft bellies and their adorable cone-shaped faces, these small, hand-sized mammals have fans all over the world. There are seventeen species of hedgehogs to be found in Asia, Africa and Europe. There are none in the Americas, although there was one in North America that's now extinct. And Down Under? Sorry, no natural hedgehog cuteness for Australia, although the European species was introduced to New Zealand in the 1870s and has become a bit of a pest, so not the biggest of smiles there.

Hedgehogs get their name from their foraging behaviour. The 'hedge' part comes from the way that they snuffle and shuffle their way through the hedgerows and undergrowth, where they hunt for snails, slugs, worms, insects and even snakes. As they trundle about, they let out little pig-like grunting sounds, hence the 'hog' part of the name. As a child growing up in Britain, I was always on the lookout for this nation's favourite. They're elusive, but regular visitors of gardens, where they make their nests in small hollows and under leaf litter. I'd always hoped to catch a glimpse of one as it searched for food or, more excitingly, as it rolled up into a spiky ball to defend itself, but I soon discovered they tend to sleep during the day and only come out at night. In fact, sleeping is something hedgehogs do extremely well. In Britain, hedgehogs hibernate throughout the winter, only appearing when temperatures begin to rise. There are, however, some species of desert hedgehog, native to Africa and the Middle East, which go into hibernation throughout the colder months, as expected, but also aestivate when temperatures get too hot.

With their sharp quills, hedgehogs are often mistaken for members of the porcupine family, but porcupines are rodents; hedgehogs are more closely related to shrews. One of the main differences is that porcupine quills come off if you get jabbed by them, while the quills of a hedgehog stay attached to the body. A third spiky mammal is the echidna, a genus of monotreme living in Australia and New Guinea. The three are a wonderful example of 'converging evolution', in which Mother Nature solves a problem, in this case protection, with the same solution – quills.

Throughout history, people have been fascinated by these little creatures, and in many countries they've even become part of legend and folklore. Down the ages, there have been some rather tall tales told about hedgehogs. In medieval Britain, it was believed that, under the cover of darkness, hedgehogs would suckle milk from the udders of cows, which led to many of them being hunted and killed. In reality, hedgehogs can't even digest the sugars that are in milk. Hedgehogs were also accused of stealing chickens' eggs but, again, even though they will eat raw egg, their mouths can't open wide enough to break through the shells. There was also a longstanding and strange belief that hedgehogs use their quills to carry food. This gem of a story arose in the first century CE, when the Roman naturalist and philosopher Gaius Plinius Secundus – aka Pliny the Elder – wrote how hedgehogs would climb apple trees, knock the fruit to the ground, then roll onto the apples to impale them on their quills, so they could carry them off and store them for winter. Bizarre though this story might seem, even the most famous naturalist of them all, Charles Darwin, once recounted a story told to him about hedgehogs in Spain: that they were 'trotting along with at least a dozen' strawberries on the ends of their quills. Just for the record, no, that doesn't happen either.

The hedgehog's quills – even if they don't carry strawberries – are a remarkable feat of natural engineering. If a hedgehog ever feels threatened, it immediately rolls into a spiky ball, which makes it painful to touch and very unappetising for any predator trying to eat it. Even more amazingly, the quills actually prevent the hedgehog from injuring itself. Pliny the Elder was wrong about hedgehogs carrying apples, but one thing he did get right was that hedgehogs are pretty good at climbing trees. Of which I had no idea! What they're less good at is coming back down again, and this is where their quills come in handy. If a hedgehog gets stuck up a tree, it just rolls into a ball and... drops to the ground. It's the job of the quills to soften the fall and prevent the hedgehog from being injured. I was genuinely surprised when I first heard about this, but hedgehogs have been known to drop from a height of six metres and be right as rain.

How does this spiky cushion work? A hedgehog has up to 7,000 quills on its back. Although they can technically be described as modified hairs, these quills have a structure that's far more complex than human hair. Instead of being solid inside, each is filled with a network of hollow air chambers. This not only makes them lightweight and very strong, but it also helps prevent them from buckling or breaking when put under stress. The tips are needle sharp and mobile, and to see how this works, we need to look at the base of the quill. Below the skin, the quill is held firmly in place by a ball-shaped follicle, which is like an anchor of cells and tissue. The follicles sit deep in the muscles running along the hedgehog's back and sides. When the hedgehog comes under attack, the muscles contract, squeezing the ball-shaped follicles and drawing the quills into an upright position. Because the muscles pull quills in different directions, the result is a crisscross barrier of spikes, impenetrable to any predator looking for a tasty snack.

This defence mechanism, as we now know, also saves all those hedgehogs unlucky enough to fall out of trees. (I still can't get over the fact they get themselves up into tree branches. Love it!) The raised, crisscrossed quills form a natural cushion and, in the very moment a hedgehog hits the ground, the thin base stems of the spikes work to absorb the shock of the impact. The result is a happy hedgehog that can roll away from danger. It's this ability to protect that's inspired an ingenious natural design for a new sports helmet, which could spell the end for crippling sports concussions.

The hedgehog's amazing ability to absorb the force of an impact is something that's very much needed in the world of sport, with concussions a major problem in any

Hedgehog

Ericaceus europaeus

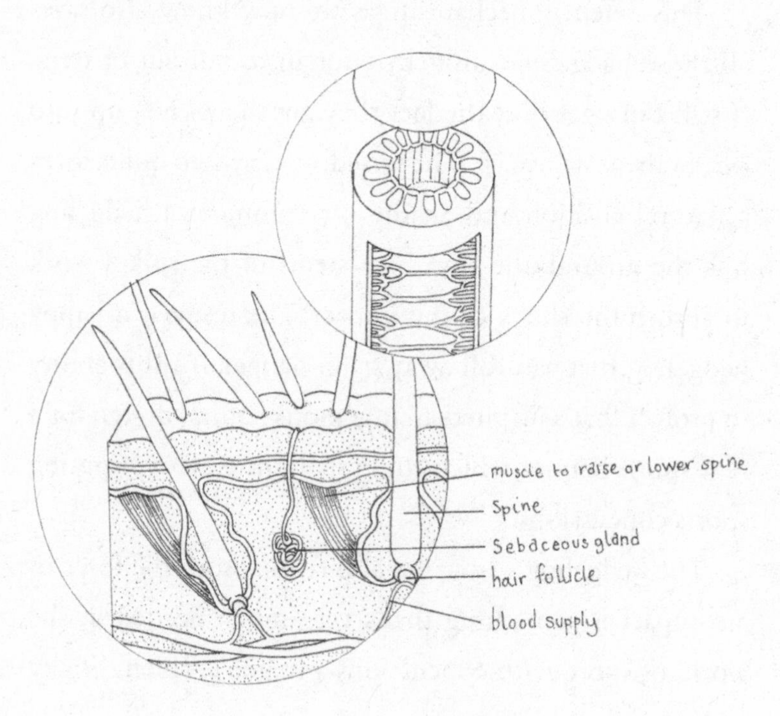

contact sport. Concussion, if you're unfamiliar with it, occurs when a blow to the body or head causes the brain to bounce and twist around inside the skull, stretching and damaging the delicate cells and structures inside it. Generally, concussions cause headaches, confusion, dizziness, nausea and issues with memory, but they can also cause long-term issues, from depression all the way through to personality changes. They've even been linked to an increased risk of developing dementia 30 years after the initial trauma. In very serious cases, a concussion can lead to death.

In recent years, however, there's been an increasing awareness of the dangers of concussion, particularly amongst younger athletes and players, whose brains are still developing. This is precisely why students from the graduate programme in biomimicry at the University of Akron, USA, decided to look to nature for a solution.

At first, they focused their attention on the hard surfaces of sheep horns and woodpecker beaks. In Chapter 4, we focused on a new type of bicycle helmet in development, inspired by a woodpecker's beak, but the students at Akron found that both the woodpecker bill and the horns of Rocky Mountain sheep tend to absorb force best from direct, head-on blows. This wouldn't work so well in sports helmets. What they wanted to develop was new helmet technology that would provide better protection against the twisting, glancing and side-on impacts that cause the most damage. The team scanned through reams of scientific papers to see if there might be another animal that was better at absorbing shock all over its body, and that's when they came across the hedgehog.

Generally speaking, sports helmets are designed to protect their wearers from the damage you get from those big hits. They have a hard outer shell, which is the part that protects the head from superficial cuts and grazes, but what really helps against concussion is the helmet's soft inner lining. For most helmets, this is made from foam or another similar material that contains air-filled chambers to provide a cushioning effect. Although these helmets work well, the team found they have a couple of drawbacks. Firstly, they become less and less effective with repeated blows. The foam lining loses its ability to spring back into shape and the air-filled chambers weaken. Secondly, the linings aren't all that good at protecting against the side-on and glancing blows that tend to cause the most damage to the brain.

Having studied the hedgehog closely at the local zoo, the team at Akron decided that their new helmet liner should look just like a hedgehog! Using a 3D printer, they constructed squares of overlapping, flexible, synthetic quills, which they attached to their liner. Each quill was printed to contain hollow air pockets, just like the inside of a real quill, to give the same strength and flexibility. The quills were then arranged in the same crisscross patterns that occur naturally on a hedgehog's back. The team named these squares 'impact protection modules', and they attached more than a dozen of them to each helmet liner. The effect was to create what looks like a 3D printed hedgehog skin within the helmet.

When struck with both direct and glancing blows, the new liner transfers the force of the impact throughout the quills, as intended, and so disperses the energy in all directions. In this way, the helmet cushions and protects the head inside, preventing that rapid, brain-rattling motion that's so dangerous to the health of athletes. And the best thing is that even when this helmet's hit over and over again, it stays just as effective, because the quills keep their shape and don't weaken.

The team is still in the testing and development stage of their helmet design, but they've had some promising results so far. After being hit at different speeds and in different spots of the helmet, initial tests show their hedgehog-inspired helmet performs better than a helmet with a foam lining. They're now experimenting with new ways of building the liners, to make it easier to produce them on an industrial scale. If all goes to plan, very soon sportsmen and sportswomen around the world might be wearing helmets that, on the inside, look like hedgehog quills. When they do take that big hit, or big tackle, they'll have the humble hedgehog to thank for keeping their brain safe and sound.

16

Packing a Punch: Mantis Shrimp

Let's face it: mantis shrimps look weird. Tucked away in burrows in tropical and subtropical seas, these strange and solitary sea creatures are stay-at-homes. They venture out only for short distances to feed. Lengthier journeys are best reserved for when moving home. Of the 450-odd species, there are a number that are active at night, with others preferring the day, and some opting for the twilight hours. A few species live in temperate seas, but most are found in the Indian and Pacific oceans, between East Africa and Hawaii.

Now, I know we're not off to a flying start on this one, but if you've never heard of a mantis shrimp before, don't let their reclusive habits fool you. They're ferocious predators, with a knockout game plan for taking out prey. Some of them spear their victims, while others club them to death with their forelimbs. They may be small, but these smash 'n' grab gangsters have one of the strongest weight-for-weight punches on the planet.

These knockout specialists are crustaceans – a diverse group of invertebrates that includes crabs, lobsters, crayfish, shrimps, prawns, barnacles and even garden woodlice. Mantis shrimps, though, stand out as among the most colourful. Take, for example, the peacock mantis shrimp – also known as the harlequin mantis shrimp. As these names suggest, it's a multicoloured dazzler, with orange legs, patches of green on its body and leopard-like spots. In fact, it comes in a stunning range of colours, from shades of pink, hues of purples and reds, to electric blues and metallic greens. But it's not just their bright colours that attract our attention.

These shrimps typically, grow to a length of about 10 centimetres, although one shrimp, caught in Florida's Indian River, has become an urban legend. It measured a staggering 46 centimetres! That's almost five times bigger than normal. Shrimpzilla just stepped into the building.

Mantis shrimps, including this giant, are covered head-to-tail in heavily armoured spines, and they have long probing antennae that detect a variety of chemical signals in the water. Most striking of all are their eyes. They're mounted on mobile stalks, and they can move independently. What's remarkable is that they can perceive depth with just one eye, while we need both eyes to do this, and they can see a world of colour beyond our wildest imagination.

To get a sense of what this world might look like, we first need to understand the difference between their eyes and ours. We humans have four types of photoreceptor cells, but mantis shrimps have an eye-watering 16 different colour photoreceptors. They can detect ultraviolet light at one end of the light spectrum, as well as infrared light at the other, along with all the colours in between. And how about sensing polarised light? Yup, add that to the list too. How does that change what the shrimp sees visually? Well, we perceive polarised light as an annoying glare, but mantis shrimps don't have this problem. Far from it: the males of some species use polarised signals on their tail and antennae to communicate with females. Since few other animals can see this polarised light, the polarised signal allows this secret liaison to take place without attracting the attention of predators. It's thought they might even use polarised light when navigating back to their home burrow.

The mantis shrimp can also detect circularly polarised light. They're the only animals on Earth known to have this ability, and it's an aspect of mantis shrimp biology that's inspiring new technologies and optical devices. One aspect of current research is the 'quarter-wave plate' action of their light-sensitive cells. These cells have the effect of rotating the plane of polarisation of light, as it travels through the cell. Human-made quarter-wave plates are found in DVD drives, and they too are required to perform this same function of converting linearly polarised light to circularly polarised light. The thing is, these devices only work well with one colour of light, but the mantis shrimp's quarter-wave plate works perfectly across the visible spectrum. Experts like Professor Nicholas Roberts at the University of Bristol, UK, remark at how this natural system 'out-performs anything we've been able to create.'

Mantis shrimps, it seems, are virtuosos of light; but they're also musical maestros capable of producing some intriguing sounds. The California mantis shrimp, for example, is known to produce its sounds by vibrating muscles underneath the cover on its back – called the carapace – which covers the rear part of the head and the first four segments of the thorax. These deep rumbles occur so fast, that they barely last a single second. Mantis shrimp will sometimes, however, produce them in successive repeats called 'rumble groups.' In addition to the rumbles, scientists have also detected 'pulsed rattling sounds,' when the carapace vibrates against another object. All this activity points to one conclusion: sound travels nearly five times faster in water than it does in air, and it's well known that marine animals like dolphins and whales use sound to communicate, so it's highly likely these sounds are being used in the same way – to ward off predators, stake claim to a territory, or attract a mate.

During the course of their life, mantis shrimps breed about 30 times, and, for some species, this is the only time that males and females meet. The female lays her eggs in her burrow, where she cleans and aerates them. In other species, in which male and females partner for life, both sexes will care for the eggs. Larvae eventually hatch out, but look and act nothing like their parents. They drift about the ocean as zooplankton. The exact number of larval stages varies tremendously between species, and they take anything from several weeks to an entire year before they transform into adults.

What really makes a mantis shrimp a mantis shrimp, though, is the possession of a lethal weapon. In fact, they get their name from the way they use it to dispatch their prey, which is very similar to the land-dwelling insect, the praying mantis. They use a pair of limbs – the second pair of thoracic appendages – to do the striking, and we can divide mantis shrimps into two teams: the spearers and the smashers. The spearers use spines on the ends of their forelimbs to pierce or slice fish and other soft prey. The smashers shatter the hard shells of snails and crabs, using extended limbs that look like a club or sledgehammer. These are called dactyl clubs.

When they unleash their punches, the smashers can strike with the acceleration of a standard .22 calibre rifle bullet, and reach speeds of up to 50 mph from a standing start. The force created by the impact is more than 1,000 times the mantis shrimp's own weight, so it's quite some punch. I've always wondered how they're able to reach these speeds. Well, it turns out that each limb acts like a spring, held in place by a latch. Whilst one muscle compresses the spring, another holds the latch in place. When it's ready, a third muscle releases the latch and... POW!

The acceleration of the club is so great that, as the strike bounces off its prey, the water literally boils, through a process called cavitation. This occurs when water that's moving really fast – in this case propelled by the mantis shrimp's club – is vaporised and turns into bubbles. As these bubbles form, they immediately implode, collapsing in on themselves to release a shedload of energy in the form of heat, light and sound. This triple-threat combo can be really destructive. It's so powerful that it can shatter the glass of an aquarium tank that's holding a mantis shrimp.

Scientists believe it's this process of cavitation that helps the smashers break apart their prey. It's hardly surprising, then, that a club, which can strike prey with

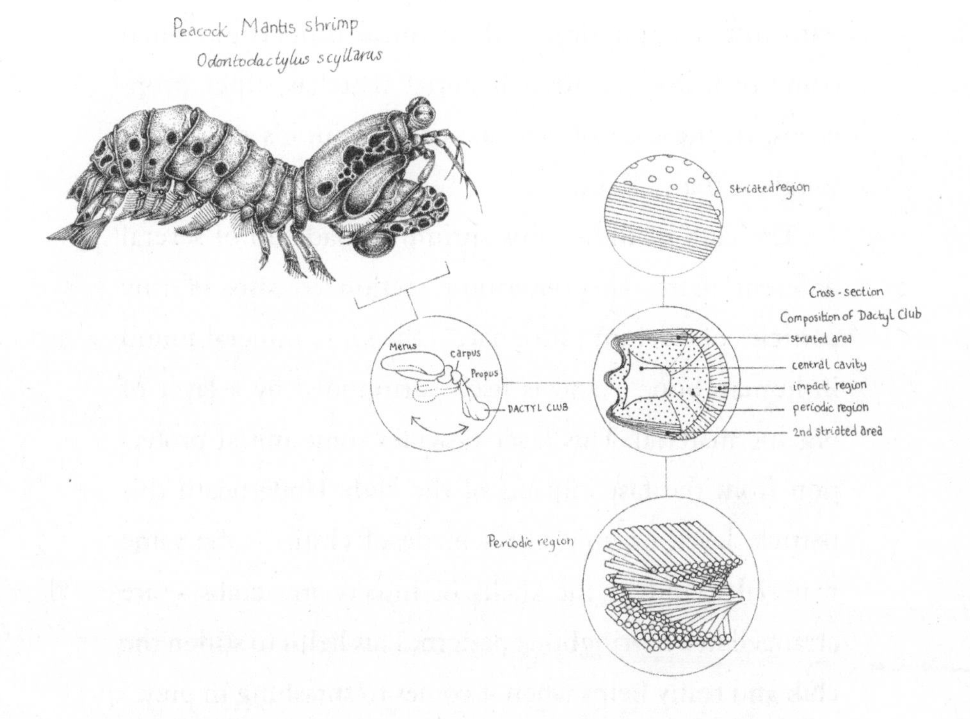

such immense force hundreds of times without breaking, has been attracting the interest of scientists, and it could change the way we manufacture materials for both the car and aerospace industries.

A team of researchers led by David Kisailus at the University of California, Irvine, and by Pablo Zavattieri at Purdue University, Indiana, has been working together to study the club-like appendage of the mantis shrimp. They've found that the bone-like mineral and organic natural fibres inside the dactyl club combine to form what's called a composite: a mixture of two materials with different physical and chemical properties. When combined, they create a material that has super properties. In the case of the mantis shrimp, it's one that's tough, stiff and light.

The club of the mantis shrimp is made up of several different parts. The outermost section consists of tiny particles of calcium phosphate – the same mineral found in human bone – and is itself surrounded by a layer of organic material. This layer provides some initial protection from the fast impacts of the club. Underneath this particle layer, organic fibres made of chitin – the same material found in the shells of insects and crabs – are arranged in a herringbone pattern. This helps to stiffen the club and really helps when it comes to smashing in prey.

Finally, the club is slammed thousands of times against the hard-shelled prey, and so it must absorb all this energy. How does it do that? David Kisailus's team found the secret is inside the club. Here, chitin fibres are arranged in what's called 'helicoidal architecture' or a flat spiral. This looks a bit like a squashed spiral staircase, where the steps circle round and round as they get higher and higher. Imagine each step is made from a layer of fibres that are lined up next to one another, like a row of pencils. Each layer of fibres is slightly rotated with respect to its neighbours.

The team discovered that when intense pressure, like a punch, is applied to this architecture, a crack starts to form. But as the crack grows it twists, and its progress is gradually slowed down by the material. This prevents what's called 'catastrophic failure'; in other words, really serious damage. In some ways, this spiral of fibre layers acts as a shock absorber. As cracks start to form, they follow the twisting pattern of the fibre staircase, rather than spreading straight out across the structure, which would cause it to fail. This differs from a material like glass, which, when dropped, has cracks rip straight through it, which then leads to catastrophic failure.

The team has since created and tested a range of materials, including carbon fibre-reinforced composites, all of which are based on the mantis shrimp's club. With the added help of cameras and digital technology they've been able to examine how their new materials react and change. They put them up against two control samples, one of which is currently used in the aerospace industry. They wanted to see which would win in a challenge of strength and resilience.

Their experiments compared the impact resistance and energy absorption of each material when they were struck, and examined their strength after impact. It was no contest. The results spoke for themselves: the helicoidal sample, based on the mantis shrimp design, scored 50 per cent better than the others. It was able to withstand high impacts and retain its strength afterwards.

The team also discovered that the spiral architecture in the mantis club is naturally designed to survive these repeated high-speed blows, by filtering out certain frequencies of vibration waves, called 'shear waves,' which are particularly damaging. Put it this way: think of the mantis shrimp's punch as a parcel of energy. If you can divert some of that energy on impact, then the punch will still be effective on your target, but won't be quite so destructive to you. The team then started to think about how this knowledge could be put into practice in the development of new composite materials that are able to filter out certain stress waves, and stop them from damaging the material in question.

In 2019, Helicoid Industries Inc. was founded to commercialise the use of helicoid architecture in composite materials. They want to apply what they've learnt to drive innovation amongst manufacturers of wind turbines, sporting goods and auto parts, all of which use components with composite materials that need to be light, strong and resistant to high impacts. So, who knows, one day we could see lighter, stronger planes or wind turbines on the horizon, which have the mighty club of the mantis shrimp to thank.

17

Snake: Search and Rescue

Snake! The very word incites fear into the hearts of many of us, and no wonder: the World Health Organization reports that up to 5.4 million people are bitten by snakes each year. That's a lot of bites, although only 2.5 per cent of those are bites by venomous snakes that result in fatalities. Needless to say, filmmakers have loved playing on our fears with their horrendously moreish snake horror movies, but it was indeed a nervousness that I had to overcome while on a search for prairie rattlesnakes.

These snakes are venomous pit vipers, so called because of the heat-sensing pit located between the eye and nostril on each side of their head. Whilst not generally aggressive, they will defend themselves vigorously, with that rattling tail a clear warning to any intruder. Pushed into a stand-off, this snake resorts to its deadly weapon of choice: needle-sharp fangs with the ability to deliver a dose of venom lethal enough to kill a human. And it's these adorable creatures that people living in southwest Canada, the western United States, and northern Mexico have as their neighbours.

With all that in mind, I did wonder if tracking one down for its venom was simply asking for trouble. Fortunately, I wasn't going it alone. My guide was snake expert Steve Mackessy, Professor of Biology at the University of North Colorado, USA, who assured me that, equipped with a trusty snake hook and a pair of gaiter coverings to protect our legs, we 'should' be OK.

The vast prairielands of Colorado made for the perfect location to track down our target. We were headed for a site that Steve had down as a nirvana for rattlesnakes: a large corrugated metal storm drain, running under a raised railroad. It was the ideal spot for snakes looking to charge up their bodies, as the heat of the sun beat down on the metal tracks, warming our cold-blooded friends in the process. A freight train rumbled past, and a flush of adrenalin ran through my body. It felt like we were suddenly living out one of the scenes from the American drama series *Breaking Bad*, wandering the desert-like landscape, preparing for mad science experiments of our own.

In all the excitement, we had to take great care. The rattlesnake is a master of disguise, a highly adept ambush hunter that's evolved the perfect strategy to capture prey. Using its flicking forked tongue to 'taste' the air, it searches for well-worn animal trails. The snake's tongue is so sensitive that it can even work out which way the most recent trailblazer was headed. It then lies in wait for its return. Its prey includes any number of rodents, from prairie dogs to gophers, as well as birds, lizards and other small mammals, like squirrels and rabbits.

Curled up into a tight S-formation, ready to strike, it uses its pit organ to detect infrared radiation, in the form of heat emanating from the body of an approaching warm-blooded animal. To the snake, the prey essentially glows, like a beacon in the night. When close enough, the rattlesnake unleashes a lightning-fast strike, momentarily sinking its fangs into its prey to pump it with venom, only to let go just as quickly. This strike and release tactic reduces the chances of the snake's sharp, yet fragile fangs, from becoming damaged, as well as avoiding any comeback from the prey.

Rattlesnake venom is a deadly little cocktail of toxins, which essentially destroy and liquefy living tissue, such as skin and muscle so, clearly, I wanted to avoid being bitten by those fangs at all costs. In the end, it was the way that rattlesnakes move that really caught my attention. I'd started out the search for these snakes with Steve quite calmly, and, believe it or not, even lay on the ground about two metres from a fully grown rattlesnake to take a closer look at the brown-checked pattern of its skin and the rattle at the end of its tail. It was, I might add, very early in the morning. The day was still cool, with the sun barely over the horizon, and so the snakes, being cold-blooded, were relatively slow movers at this stage.

Returning after lunch, it was a completely different story. Walking back towards the storm drain, I let my guard down for but a moment, and that's when one of the biggest rattlesnakes I'd ever seen came racing across the path in front of us. I froze, as if Medusa herself had me locked in a stony gaze. My director, though, had other ideas. Virtually jumping out of his skin, he instinctively grabbed a hold of me, like some kind of human shield. I didn't know whether to laugh or cry, but I can tell you that my skin was alive with goose bumps.

Despite my trepidation, I must admit that I have a lot of respect for this animal that, with no limbs, can move with such ease. It's a feeling I'm sure is shared by herpetologists and snake lovers across the world. Today might see you added to that list, because it's a snake that's proved to be the source of inspiration for a new search-and-rescue robot that, in the future, could save countless lives.

There are around 3,700 species of snake that we know of, and like them or loathe them, they're found just about everywhere, with the exception of Antarctica, Iceland, Ireland, Greenland and New Zealand. Some 600 of these species are venomous, the Asian king cobra being the longest, with a length of more than five metres, and Australia's inland taipan being the most venomous. Of the non-venomous snakes the seven-metre reticulated python of southern Asia is the world's longest, with the heaviest being South America's green anaconda, which can weigh up to a staggering 100 kilograms. While the king cobra, much like rattlesnakes, kills prey using venom, the two giants dispatch their prey by constricting them until they die.

Whether they kill with a venomous strike, by squeezing, or by eating prey alive, like the harmless garter snake, nearly all swallow their food whole. Many snakes can eat animals up to three times bigger than their head is wide; as you can imagine, this can be quite a mouthful at times. This ability all comes down to the fact that their jawbones are loosely connected, which means they're more flexible than ours. Once inside the snake's mouth, the prey is held in place by backward-facing teeth which prevent it from escaping or falling out.

It's the ability of snakes to move so effortlessly across a complex variety of landscapes that's captured the interest of a team of engineers led by Chen Li, assistant professor of Mechanical Engineering at Johns Hopkins University,

USA. Could a snake, they wondered, be the inspiration for life-saving, search-and-rescue robots of the future?

Chen Li's team set themselves the challenge of designing a robot that could move through narrow gaps and over the type of rubble and debris you'd expect to find in the aftermath of an earthquake, for example. A snake-like robot seemed the perfect answer. It doesn't really take up that much space, it'd be able to slide through small gaps, and get up and over large objects. In the past, studies of terrestrial snakes have mostly focused on their movement across flat surfaces, but little was known about how snakes move over what's referred to as '3D terrain.' We're talking about uneven surfaces and objects like fallen trees and rocky boulders. It's far more challenging to move across these sorts of surfaces than it is on flat ones. You'll know this yourself, if you've ever gone for a run through the uneven terrain of a forest. You're far more likely to wobble and slip, or even fall over, than if you're running along a flat paved street. On uneven ground, you're less stable. The same goes for the robot. To draw inspiration for their snake robot, the team had to study a real snake species that's particularly well adapted to moving across rough, rocky ground.

The snake they chose to study was the variable kingsnake, a species found in a variety of landscapes on the plateaus of the northeast Mexican state of Tamaulipas. Adults can grow up to 80 centimetres long and, luckily for Chen Li's team, are non-venomous. They kill by constriction and, as their common name suggests, their colour pattern varies quite a bit. There are four different colour combinations: leonis, buckskin, milk and black, and they can all appear from a single clutch of eggs, although the black colour combination is quite rare. Oh, and if you're wondering where their common name 'kingsnake' comes from, it's down to their habit of eating other snakes. They'll even eat young rattlesnakes, a behaviour known as ophiophagy. In the wild, they encounter all sorts of terrain, so you could consider them masters of movement. A bonus for researchers was that they also thrive in captivity.

The team started out by running a series of experiments to see how the snakes bend their bodies when they're faced with barriers. They presented them with steps of different heights – five or ten centimetres – and with different surfaces of either rough cloth or smooth paper. They watched three snakes, and the experiment was repeated, so that each snake climbed ten times on each step height and each step covering. The results were revealing.

To climb a step, the snake in effect divides its body into three sections: the front and rear sections stay close to the surface and wriggle back and forth on the flat horizontal steps, while the middle section of the body, stretched between the steps, is suspended in the air to bridge the large gap. The wriggling portions, the team noticed, provided stability, acting to support the snake, and to stop it from tipping over. When they made the steps taller and

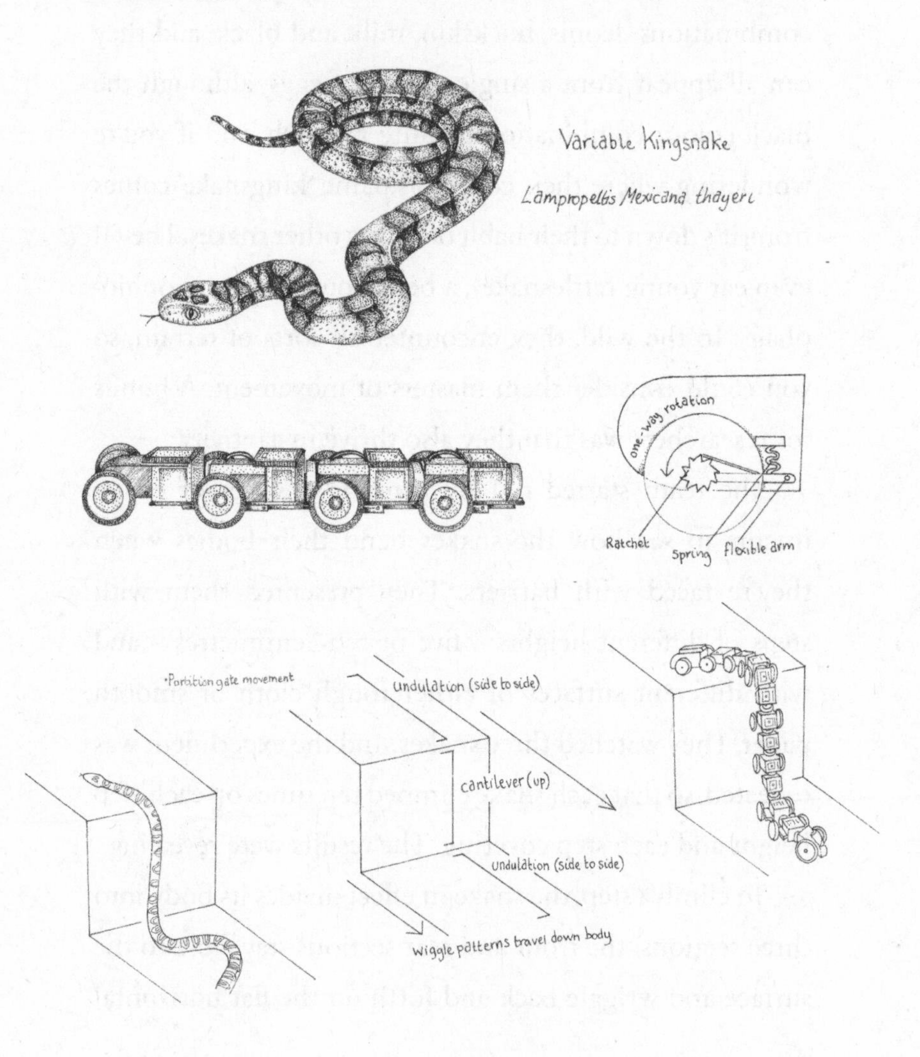

more slippery, the snakes would move more slowly, and would wriggle their front and rear body sections less. In this way, they maintained stability and kept their balance, so they didn't fall over.

After watching videos of the snakes climbing up the steps, Qiyuan Fu, a graduate student at Li's lab, built a robot to copy the snakes' movements. Its main body is a metre long, weighs just over two kilograms, and looks a bit like a toy train with lots of carriages. In total, it has 19 carriages or segments. Special joints between them enable each segment to rock, either up and down, or left and right, which provides the robot with the flexibility to climb over the steps. A pair of one-way wheels is attached to each segment. The wheels only unlock when they're rotating forwards, and they lock again when they rotate backwards. This means there's a bit of friction when the robot moves forward, and a lot more friction if it tries to go backwards or sideways. This setup is very similar, in some ways, to how a snake's scales work, and help move the robot forwards.

In early trials, the snake robot had some difficulty remaining stable or balanced on the large steps, and often wobbled and flipped over, or got stuck on the steps. The real snake, by contrast, is always stable. So, how does it manage this? You'll recall that when the snake was trying to get up the step, it kept the middle section of its body straight in the air. This is called cantilevering. Occasionally, it leans against the vertical surface of the step to help keep its balance. The snake spreads the wriggling section, which gives it a large 'footprint' on any given surface, and this prevents it from rolling over. Most importantly, unlike the robot's rigid wheels, which frequently lose contact with the flat surface due to the way the robot wobbles, the snake's soft body stays in contact with the flat surface all the time.

The team decided to improve the stability of their robot snake by adding a car-like suspension. This was basically a spring between the body and each wheel, which enabled the wheels to maintain good contact with the surface it was on, even when the body section wobbled. The springs acted as a suspension system, so when each wheel is pushed against the surface, its spring would compress in order for most wheels to maintain contact. With the new suspension system, the snake robot was less wobbly, and could climb steps as high as 38 per cent of its body length, all with a near 100 per cent success rate, double what it was before adding the suspension.

Compared to snake robots from other studies, Li's robot had the disadvantage of needing more electricity to account for the suspension system, but it made up for this by the sheer fact that it was speedier than all but one of the other robot alternatives, which itself was less stable. Li's robot snake even came close to matching the speed of the real variable kingsnake.

The team plans to continue to work on their 'snake-bot', as they call it, and make it even better at tackling more complex 3D terrain or surfaces with multiple obstacles. Their hope for the future is to see their snake robots climbing up mountains, through earthquake rubble, or even across Martian rocks. Fitted with remote sensors and cameras, they'll be more than primed to help us with a whole range of tasks, be it search and rescue, environmental monitoring or planetary exploration. Of course, quite how you'd react if you were buried under rubble and saw a snake robot heading towards you is another matter!

18

Natural Architects and Artists: Butterflies

When a butterfly flutters by, you can't help but notice its erratic flight and flashes of colour, and the blue morpho butterflies of South and Central America are among the most beautiful in the world. With a wingspan up to 20 centimetres, they can be bigger than your hand. It's not only their size that's so striking: their brilliant blue wings appear to shimmer. Performing a 'flash and dazzle' wing display, males use their bright colours to intimidate rivals that fly into their territory, attract the opposite sex or confuse predators when threatened. Females are not blue and are a little less vivid than males, but are nevertheless impressive in their own right, with wings in various shades of brown, yellow and black.

The males, though, are so impressive that these butterflies have special significance to people living alongside them. They are considered a powerful representation of the soul and of spiritual transformation. The shimmering blue colour is thought to symbolise healing, and people in Costa Rica often make a wish when they spot the butterflies. These stunning insects, however, may have another significance: they could also hold the key to transforming our homes.

Blue morpho butterflies are by no means the only iridescent insects. The most famous, perhaps, are the jewel beetles. Seeing them in the flesh, you'd be forgiven for questioning whether these beetles were even real. As creatures of sparkling beauty, they really are nature's jewels, and some species have been highly prized since the seventeenth century, with their wing cases stitched into ceremonial costumes, headdresses, decorative fabrics, jewellery and artwork. Sometimes we're not just talking about one or two beetles: a team led by Belgian artist and director Jan Fabre used more than one-and-a-half million beetle wing cases in the redecoration of the spectacular Hall of Mirrors in the Royal Palace of Brussels. Each is arranged on the ceiling, following a detailed sketch by the artist. It took Fabre's assistants four months to carefully glue them in place in what he called *Heaven of Delight*, and, of course, an awful lot of beetles died as a result.

Iridescence is not confined to insects. Birds, such as magpies and hummingbirds, have iridescent feathers and, like the insects, they have been highly valued. In the past, iridescent bird feathers were used to make royal cloaks in a number of Polynesian cultures. Similarly, the reflective coating known as mother of pearl, which lines the inside surface of mollusc shells, has been used in jewellery. Artists from Japan to the Ottoman Empire have created carved decorations as a show of wealth and status. It even featured in Aztec culture.

The term iridescence comes from the Latin and Greek word *iris*, meaning 'rainbow', and also refers to the Greek goddess Iris, who is said to be the personification of the rainbow, and acted as a messenger to the gods. Due to their changing nature and the range of shades they produce, iridescent colours are often described as rainbow-like, shimmering, metallic or sparkling. They even occur in non-living things that are visible to us every day. You've probably seen this effect in soap bubbles or even on small pools in the road with oil on the water.

But, returning to the blue morpho butterfly, I can reveal that, while the upper surface of the male butterfly's wings displays a brilliant and iridescent cobalt-blue colour, they actually contain no blue pigment. It's a type of butterfly that beautifully demonstrates what's called 'structural colour', and this has fascinated scientists, who are now looking at just how these butterflies produce this shimmering blue effect in order to create pigment-free paints and textiles.

The colour is the direct result of light reflecting from the tiny scales that cover a butterfly's wing. Under the microscope, they resemble rows of overlapping tiles, but, to see how the colour develops on them, a research group led by Nipam Patel at the University of California in Berkeley, USA, have had to look inside the chrysalis, where the adult butterfly and its wing colours are developing. They found a way to remove the wings from inside the pupal case and then grow them in a petri dish. Like a photograph developing over time, colour and patterns slowly emerged on the white wings, as each scale was transformed. The key structure is tiny ridges on the surface of the scales. These cause light waves to reflect and interfere with one another, to make some colours brighter and others darker. When light hits the ridges, something called 'constructive interference' happens. This occurs because, instead of having smooth sides, the ridges are indented, like a double-edged comb. The spacing within the ridges – or between the teeth of the comb – determines which wavelengths of light will be reinforced and which will be cancelled out. This is why our eyes see this shimmering blue effect.

Researchers like Patel's team see the potential in structural colour for developing new pigment-free paints and

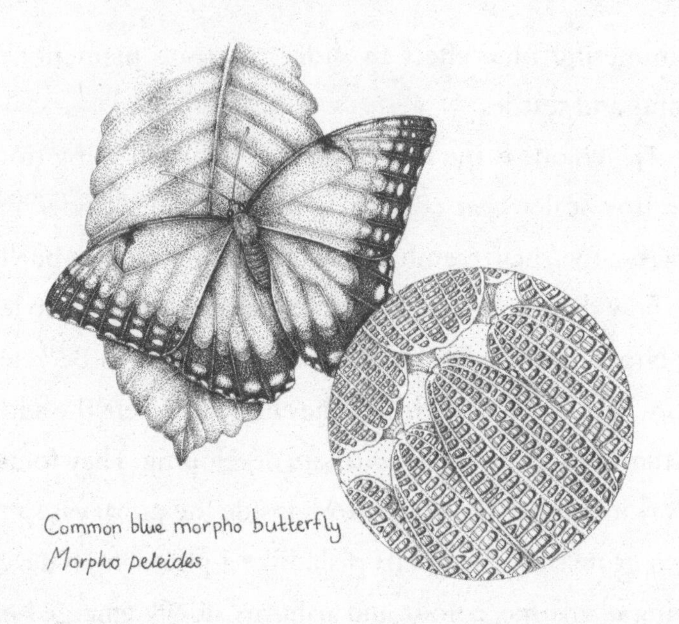
Common blue morpho butterfly
Morpho peleides

fabrics. Imagine! Cars whose colour is determined not by traditional chemical paints, but from the structure of the sheet metal they're made of... Well, imagine no more.

A new range of paints, designed primarily for cars, contains a light-interference pigment, so the paint changes colour depending on the light source and viewing angle, much like the butterfly's wing. The effect is achieved by interfering with the reflection and refraction of light from the painted object's surface. The paint contains tiny synthetic flakes, about one micrometre thick. They're made from aluminium coated with glass-like magnesium fluoride, embedded in semi-translucent chromium. The aluminium and chrome give the paint a vibrant metallic sparkle, while the glass-like coating acts as a refracting prism, changing the apparent colour of the surface as the observer moves. Similar paints have also been used as a substitute for optically variable ink in anti-counterfeiting currency. The technique is not confined to paint – textiles might benefit from structural colour too.

In the Netherlands, Teijin Fibre Corporation has created a fibre made of 61 nylon and polyester layers capable of producing basic colours such as blue, green and red without the use of pigments. Once again, this was based on the wings of morpho butterflies. The process reveals that it's possible to produce a range of colours, depending on the intensity and angle of light. Traditionally, pigments are used to introduce colour to textiles and clothing, but this can be hazardous. They can be the cause of serious environmental pollution, because non-biodegradable dyes can be spewed into rivers in dye factory effluents. Dyeing can also give rise to air pollutants; in fact, in many places in the world, the dyeing industry has left cities with severe pollution of both surface and ground water. As no actual dye is involved in producing these new and innovative colour-shifting fabrics and paints, it reduces the need for the large amounts of water and energy that are associated with conventional dyeing. Another advantage of structural colour is that it doesn't fade.

SOLAR-POWERED ROSE BUTTERFLY

Another spectacular butterfly is the common rose butterfly of southern Asia. It belongs to a family of butterflies known as swallowtails, recognised by the trailing edge of the hind wings that are forked in a way that resembles a swallow's tail. The name 'rose' comes from its bright crimson body and crimson markings, which are the colour of a red rose. It also has striking white spots on the underside of its wings, and these bright splashes of colour are an anti-predator device. They work by either startling predators or by deflecting attacks away from the butterfly's vital body parts.

What's particularly striking about the rose butterfly, though, is the dominant colour on the upper side of its wings – a deep, dark and velvety black. The amount of blackness varies between males and females, the males being the darker, but these stunning wings are not just for show. Scientists have discovered that the rose butterfly's wings have evolved to be quite brilliant at collecting sunlight, and it's this ability that optical engineers are copying to create a new type of thin film solar cell.

Butterflies, like all insects, are cold-blooded. This means they can't generate all the heat and energy they need from their own bodies. Instead, they have to rely on the heat from their environment to warm themselves up to their optimum operating temperature, and that's about 30°C. You've probably seen butterflies resting with their wings outstretched in a patch of sunlight. This isn't just because sitting in the sun is a pleasant thing to do. The butterfly must harvest enough solar energy to fly. Basking, with their wings outreached to soak up the sun's rays, is the best way for butterflies to raise their body temperature, and dark colours absorb more heat and light than light colours which reflect it.

With this in mind, Radwanual Siddique – an optical engineer from the California Institute of Technology, USA, working also with the Karlsruhe Institute of Technology, Germany – was intrigued by the way microscopic structures on butterfly wings led to their shimmering iridescence, and wondered whether there was something here that could improve solar cells. He first studied the blue morpho, but on a visit to a butterfly nursery in Mannheim, Germany, Siddique found himself captivated by the rose butterfly. Fascinated by the velvety texture of its black wings, he asked the nursery if he could collect a few samples to examine under an electron microscope to understand exactly how they work.

He saw that the surface of the rose butterfly wings were made up of the thousands of tiny scales as he'd seen on other butterflies, but closer examination revealed something quite remarkable. On each of these scales were more intricate structures made of ridges, which crisscrossed each other to form a sort of lattice. This contained lots of microscopic holes of different sizes arranged in a disordered pattern. The lattice was built from chitin, the hard material found on the exoskeletons of insects. It also contained melanin, which is what gives our hair and skin its colour, and absorbs light. Turns out, it was the melanin that provided the butterfly wings with its beautiful dark colour.

Siddique realised this microscopic lattice structure was improving the way the butterfly's wings were absorbing light. It was doing this because the ridges and holes were scattering the light that hit the wings, and reflecting it back into the tiny scales to be absorbed by the melanin. So, it wasn't just that the butterfly wings were absorbing light directly from the sun, they were also magnifying and enhancing this light, so they could reabsorb as much of it as possible. Siddique had discovered that the wings of the rose butterfly were not only pretty, but that they were also superefficient, super-thin light absorbers.

It was this discovery that opened up a world of exciting potential for the optical scientist, who was already looking into ways of improving solar panel technology. The most efficient types of solar panels are made from thick crystal-based cells. You've probably seen them positioned on rooftops, on buildings or in fields, and they're placed at the precise angle that would allow them to generate the most energy from the sun as it moves across the sky. This works really well when the sun is at its optimum position, but when it moves to one side, the solar panels stay where they are and so, in essence, become less and less effective. One solution to this problem, would be to mount the panels on a moving structure so they can follow the sun as it travels across the sky, but this would be an extremely expensive solution. Not only would you need to buy motors to move the panels, but you'd also need additional electricity to power them, not to mention the added maintenance costs.

There is another type of solar panel that's made up of thin film solar cells. You see these on small devices such as calculators and watches. They are much cheaper to produce and have a light-absorbing layer a thousand times thinner than the crystal-based solar cells. However, they tend to absorb less light and are far less efficient, which is why they're not used on solar panels that need to generate a lot of energy. Siddique and his team wondered if applying the rose butterfly wing lattice structure to thin solar cells could improve their efficiency. If successful, they believed it might have the potential to open up countless new possibilities to use this cheaper, lighter material. They might even be able to use their redesigned thin solar cells on large solar panels.

So the team had to find a way of making the same lattice structure that covered the rose butterfly wings in the laboratory, and the solution they found was ingenious. Instead of trying to build their own version of the microscopic lattice structure, they found a way to create it organically. They mixed together two different liquid plastics into a solution, which they then spread out onto a flat surface. As the plastics dried, they repelled each other, because they were made from different materials, and the solution naturally organised itself into the same lattice structure as on the butterfly's wings.

The team transferred their plastic lattice structure onto a solar cell and measured how much light it absorbed. They found that, compared to a conventional smooth surface, their solar cell absorbed 90 per cent more direct light than a conventional cell, and when the cell was placed next to a light source that shone at an angle, it showed an increase of a staggering 200 per cent more than if they shone the light at an angle to the conventional cell. They knew this was a really exciting development, because it meant the cells would be particularly effective in countries that wanted to use solar panels to generate electricity, but had only minimal hours of sunlight.

Siddique and his team are still in the development stage of their rose butterfly-inspired solar cells. But so far their tests have shown that they perform really well compared to other models currently on sale, plus (and perhaps most importantly) they're cheap and easy to make. They're also working on applying this technology to other optical structures, such as LED lights and biosensors. Their hope is that their invention will help develop cheap and efficient solar panels that can be used in a greater variety of situations than they are right now, which would help reduce our dependence on fossil fuels. For this remarkable innovation, we have to thank the wings of a pretty red and black butterfly.

THE BUTTERFLY HOUSE

At the time of the European Renaissance, the fifteenth-century architect Filippo Brunelleschi designed a thinner, lighter dome for the Cathedral of Florence – the Duomo di Firenze – in northern Italy, after carefully studying the strength and structure of eggshells. The story goes that, when the cathedral was built, no one knew how to design the dome. In 1419, the Wood Merchants Guild held a competition inviting architects to submit their ideas. When Brunelleschi was invited to show the others the model of his design, he flatly refused. Instead, he issued a challenge, proposing that the commission should be given to the man who could make an egg stand on its end on the table in front of them, as that man would have the skills required for the job.

After all the other architects tried and failed, Brunelleschi took an egg, tapped its end to crush it ever so slightly, and then placed it onto the table, where it stood upright. Infuriated, the other architects responded that they could have done the same thing, but Brunelleschi laughed and replied that they could also build the dome if they could see his model. Impressed by him, the judges decided to give Brunelleschi the commission – and you know what – the dome does resemble an egg slightly flattened at the top. It took some 17 years to complete and it's the largest dome ever created in brick and masonry.

In recent years, advances in digital technology and 3D printing mean that we can now create shapes and structures, some based on living things, that otherwise would've been impossible. One example is the design for a butterfly house called *Metamorphosis: Inception* by Tia Kharrat, a graduate of the University of Westminster.

Just like the cathedral dome of Filippo Brunelleschi, Tia's design also mimics the shape of an egg, this time the egg case of the white royal butterfly, an endangered species from Singapore. The spherical eggs are laid on the underside of leaves and are white with a hint of green at first, the colour disappearing after a few hours. Even though each egg is tiny – less than a millimetre in diameter – it's covered with a beautiful fractal design of pits and ridges. Fractals are really cool – they're the sort of thing you can see if you were to look at a snowflake under a microscope, where the various patterns are repeated over and over again at different scales in the same object.

Tia replicated this, using 3D software and some complicated mathematics, to create a digital representation of the egg, which she then made into a physical object using 3D printing. This resulted in a structure made of concave, domed hexagonal panels. She compares it to being inside a football, where each of the panels is perforated with holes, to allow light and air to filter though.

Tia points out, however, that biomimicry is not just about copying forms from nature; it's about taking the time to look at the designs that occur in nature and to use them to solve problems and create designs in a different way. She used her studies of the butterfly egg to inspire a different form. Her final design is not a literal copy of the egg, but rather takes the geometric rules and multiplies them in a fractal manner to design something new.

She first started work on her design as part of her master's degree, and, appropriately, the pavilion was originally designed to house a butterfly enclosure. As an aspiring conservationist, Tia also wants her work to raise awareness about the global decline in butterfly numbers, although she feels that the relationship between humans and butterflies should not only be about conservation, but a celebration too. She also believes that a structure like the butterfly-inspired enclosure could have a positive psychological impact on its inhabitants, her theory based on the fact that we are somehow subconsciously familiar with naturally occurring patterns, so we should be more relaxed when surrounded by them.

While her design was intended originally for a butterfly house, it could lend itself to other uses, such as holiday retreats and places of worship. The use of translucent materials, the spatial arrangement and the dappled light all combine to create an area where people would be more inclined to contemplate life, and even, in her own words, 'experience the first stage of their own metamorphosis'... an intriguing idea.

19

Giant Fish and Body Armour

The Amazon River! My younger mind couldn't help but marvel at its mysterious nature. I'd always imagined it to be a worthy contender for the title of 'greatest river on Earth.' In my mind, there were two challengers: the mighty Amazon and the jewel that is the Nile. Having been born to Ghanaian parents, I was satisfied with Africa claiming victory; after all, the Nile is officially the world's longest river. The biggest river by volume, though, is the Amazon – it's in a league of its own.

The river and its tributaries drain an area of about 6.9 million square kilometres, according to the US Geological Survey, and it stretches from the Andes mountain chain to the Atlantic Ocean, accounting for 38 per cent of the total area of South America. Its most distant point from the sea has been hotly debated. Many sites have featured down the centuries, but in 2007 research by Brazil's National Institute for Space Research, along with other organisations, opted for Apacheta Creek as the source of the Amazon. From here, they calculated that the distance to Marajó Bay, in the southern part of the river's mouth, is about 6,992 kilometres. Using the same technology they found the Nile to be about 6,853 kilometres, which would make it the second longest river. So the Amazon is top, but for how long?

When the Amazon reaches the sea, a mind-blowing 219,000 cubic metres per second of freshwater discharges into the Atlantic, which represents more than the rest of the world's seven largest rivers combined. The delta is 320 kilometres wide, and the brown waters containing the river's sediments can be detected 100 kilometres out to sea.

Little more than 190 kilometres from the Pacific Ocean to the west, into which the Amazon once flowed, the headwaters plunge in great cascades from the eastern slopes of the mountains, especially when the snows melt in spring, but downstream from there the river levels out. It descends no more than 1.5 centimetres for every kilometre over several thousand kilometres, so the waters meander relatively slowly, forming numerous oxbow lakes that are havens for wildlife.

In the dry season, the main Amazon River can be 4–5 kilometres wide in places, but in the wet season this increases to 50 kilometres, its waters drowning the surrounding rainforest, an area known as the *várzea* or 'flooded forest', which includes some of the most productive parts of the Amazon Basin. Here, river dolphins, manatees, stingrays and other extraordinary fish swim amongst the treetops, some feeding on fruit and nuts. One of the biggest fish – in fact, one of the largest freshwater fish on Earth – is the arapaima or pirarucu. The largest ever recorded was 3.07 metres from head to tail and weighed 200 kilograms, but even longer ones – up to 4.5 metres long – have been spotted, their length not verified.

Sadly, the only glimpse I've ever had of the arapaima was in a restaurant aquarium in the Pantanal, an area of wetland to the southwest of the Amazon Basin – and no, don't worry, I didn't order it for dinner. Even there, I could see that this Amazonian giant was something special. Looking at it was like taking a glimpse back through time, for it is a member of a family of fishes, known as 'bonytongues', which have been described as 'living fossils' and even 'dinosaur fish.' It's a species that has been on this Earth little changed for at least 13 million years (the age of a fossil found in Colombia), making the arapaima one of the most ancient freshwater fish in the world. Its green, broad head is unmistakable, along with its long, streamlined body and a massive flecked-red tail but, like many animals in the region, it is threatened with extinction. As I found from my trip to the restaurant, arapaimas are considered pretty tasty and are sometimes referred to as the 'cod of the Amazon.' For centuries, local people have fished them. They dry and salt them to preserve their flesh.

One of the arapaima's more unusual adaptations is that it doesn't use its gills to breathe. Odd that – a fish that doesn't use its gills! So how does it survive? It takes in air through a modified swim bladder, which acts as a primitive lung. This does, however, mean that the fish has to swim to the surface every 10–20 minutes to take a breath. If you're ever close to an arapaima, you'll know it has surfaced because it makes a distinctive noisy gulp, which sounds a bit like a strange cough. It can be heard from quite some distance away. It has evolved this way of breathing because the floodwater lakes in which it lives are generally low in oxygen. In the dry season, as the floodwaters drain away, the lakes become clogged with decomposing vegetation, so oxygen is used up rapidly. Thanks to its air-breathing lung, though, the arapaima is able to thrive in these waters, feasting on fish that find themselves cut off from the river.

Now it may sound like arapaimas have worked out how to live an easy life, but they're not the only predator around here. The lakes also attract shoals of another Amazonian fish, which likes to lay its eggs among the rotting vegetation. It's one with an infamous and brutal bite – it's the piranha.

If you're anything like me, you'll be familiar with piranhas. My cousins and I became obsessed with them. I'd squirm at, but also relish, the pinnacle of an action movie, when 'the baddie' would dispose of his or her enemies by throwing them into a tank of piranhas, and then watch with glee as the feeding frenzy began. Between my cousins and me, the debate would rage on for weeks about exactly how long it would take for a person to be eaten alive. The reality is that these fish are not as dangerous to humans as is made out. They might take a chunk out of a finger or toe, but attacks are rarely fatal, although, recently, there have been a few exceptions: in 2010, a drunk youth, who went swimming, was killed in Bolivia; in 2012, a young girl was attacked and killed by a shoal of red-bellied piranhas, and, in 2015, another died when her grandmother's boat capsized – both attacks were in Brazil.

In the language of the Brazilian Tupi people, the word *piranha* translates as 'tooth fish,' and for good reason. Piranhas have rows of sharp triangular teeth that come together like a pair of crimping scissors. Their jaws are powered by huge muscles that take up most of the space in their heads, which gives them an extremely strong bite. When they grab hold of their prey, they'll start thrashing their tales back and forth to help them prize flesh away from bone, which means, if they're eating a small animal or fish, they can reduce it to a skeleton within seconds.

The piranhas – of which there are several species – living in the Amazon's floodwater lakes tend to feed during the day. As the water level drops and food becomes scarce, they become much more aggressive, eating anything they can find. This is when they're more dangerous. They've even been known to grab birds, especially young egrets that have fallen out of nests located over water. This is why many other species feed mainly at night, in order to best avoid their bite, but the arapaima, with its need to take a gulp of air every 10–20 minutes, doesn't have this luxury. Instead, the giant fish has developed an ingenious protection system.

The fish's body is covered in layers of grey-green, armour-like scales, tightly layered, one on top of another. Even when grasped by the piranhas' powerful jaws, these scales don't tear or crack. The arapaima has evolved its own body armour, and what's remarkable about it is, unlike the type worn by soldiers, the fish-scale armour is super flexible. This means that, although arapaimas are tough, they can also move easily in the water and quickly swim away from danger if they need to. Unsurprisingly, it's caught the attention of scientists with an interest in military body armour.

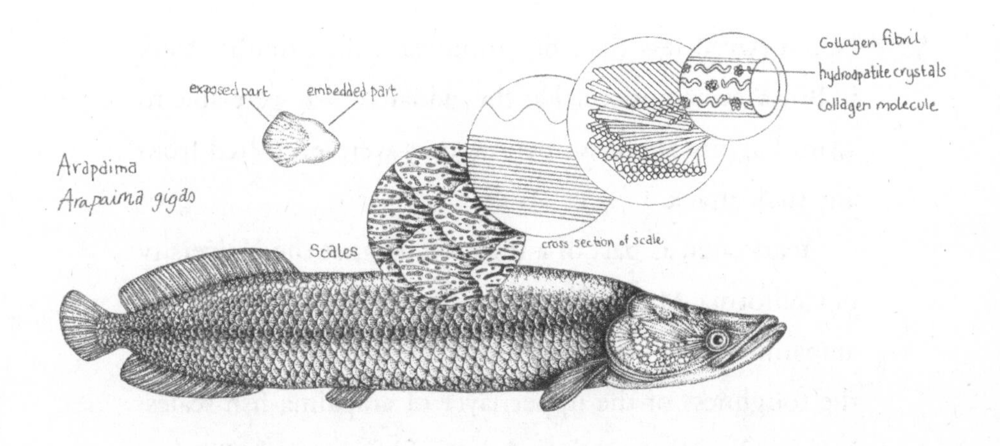

American materials scientist and engineer Professor Marc A. Meyers at the University of California, San Diego, grew up in Brazil, and tales of the rainforest and the creatures that lived there fascinate him. As a young man, he decided to visit the Amazon Basin with his friends. They travelled for two gruelling days on the back of a truck before they reached the river. When they arrived, hot and tired, the group wanted to jump straight in the water to cool off but, just in the nick of time, they stopped themselves. They realised the river was teeming with piranha. They'd all heard the stories of people who had fallen off boats, only to be been eaten alive, and weren't keen to repeat the experience. He also saw how local fishermen used chunks of meat as bait to catch an arapaima. As the arapaima surfaced, they would throw the bait just in front of its mouth, but if the big fish didn't take it, the bait survived less than five minutes – the piranhas took it. But Meyers noticed that the arapaima fish were able to swim happily about; somehow, they were protected from any such attack.

Years later, as part of a research team at the University of California, Meyers began to investigate exactly why the arapaima was so piranha-resistant. He was already aware of the toughness of the upper layer of arapaima fish scales: he'd seen local women use them to file their nails. But just how tough were they? And did the scales have some other property that helped to protect the fish?

Meyers asked a friend in Brazil to send him a box of scales, so he and his team could examine them. He first cemented a piranha tooth to a machine that delivered a 'bite' to the scales. The tooth didn't just ricochet off the armour, but shattered. This was something worth closer investigation. Using the scanning electron microscope, they found that the scales, which can be up to ten centimetres long, were far thicker than scales from other large fish. When it came to studying the outer layer – this is the part of the scale that was being used as a nail file – they noticed it had a similar composition to that of human bone, but it contained more mineral deposits and was much harder. It was only when they observed the entire scale under the high-powered microscope, however, that they understood why it was so remarkable. Instead of being smooth, the scientists observed that the tough, outer layer of the arapaima scale contained ridges and bumps, a bit like furrows in a field. These were arranged in a zigzag pattern. This meant that although this layer was extremely strong, it had some flexibility.

The discovery of what was happening in the scale's silvery-red inner layers was even more exciting. Here, the team found long strands of the protein collagen. This is the same type of protein that makes up connective tissues in cartilage, bones, tendons, ligaments and skin. These strands were bundled together to make tiny fibrils with a thickness of only a micrometre – that's barely one hundredth of the width of a strand of human hair – and the fibrils were arranged in sheets, which lay one on top of the other, each slightly rotated against the one above. The result was a spiral staircase-like construction, known as a Bouligand-type structure. This showed that, even if a sharp tooth were able to break through the hard, outer layer of the scale, it wouldn't be able to crack or tear the complex collagen structure beneath.

To test this, Meyer made a crack in one scale and soaked it in water for 48 hours. He then slowly pulled its edges apart, while adding pressure to a central point. As he did, he saw that part of the hard outer layer of the scale expanded, cracked and then gradually peeled off, but the scale underneath remained intact. On closer inspection, he saw that the collagen fibres had lost their neat spiral-staircase-like structure and, instead, had lined up to face the direction of the force. This meant the fibres were providing the maximum resistance possible. Even under intense pressure, the inner collagen layers – although they moved and deformed – didn't break. It's this property that makes the arapaima scale among the toughest and most flexible biological materials on the planet.

This discovery has the potential to inspire a new type of body armour. Currently, some bullets and knives can penetrate the plastic polymers used for flexible body armour. A new synthetic material that is hard on the outside but which bends, like the inner and outer layers of the arapaima scales, while remaining strong, would stop all bullets and other weapons. Marc Meyer's team in California is currently looking at 3D printers to produce this... and that's not all.

The inner Bouligand-like structure of the arapaima scales could be used in aerospace design, to make protective cases for jet turbines and space vehicles. It could even be used in submarines to help them hide from their opponents, because this new material also has the awesome ability to absorb the energy given off by the craft, making it undetectable by sonar. In fact, the discovery of how the Bouligand structure works is sparking a revolution in materials engineering – all this inspired by the scales that have protected arapaima in the piranha-infested waters of the Amazon for millions of years. In the not too distant future, soldiers and service personnel all around the world may have a prehistoric, coughing fish to thank for keeping them safe.

20

Cows and Eco-Friendly Sewage

One of the very first animals many of us learn about as we're growing up is the cow. As a young boy, I remember looking at picture books and singing songs about cows that lived on farms. There was even an old rhyme about a cow that jumped over the moon. It's no coincidence, then, that children still sing songs like this, because cows and humans have an important, and, perhaps, a far more intertwined relationship than we might realise.

Cows have been part of human life for thousands of years. It's believed that cows were first farmed as livestock around 10,500 years ago, with all the cows on western farms today descended from a herd of aurochs – a species of ferocious wild cattle that lived in Eurasia and which is now extinct – that some brave farmer domesticated in what is now Iran. Way before that, our ancestors hunted the aurochs and other wild cattle. They appear in 17,000-year-old cave paintings in Lascaux, southern France, and some of the oldest figurative cave paintings ever found – dated more than 40,000 years ago on cave walls in Borneo – show humans alongside cow-like animals.

In early cultures, cows were regarded as sacred and worshipped because their crescent-shaped horns resemble the different phases of the moon. Many great ancient civilisations had some form of cow god, and even today, Hindus consider the cow to be sacred. Others put it to work. When humans began cultivating grain, cows were among the first animals used to work the land. Owning cows has always been a sign of wealth and prestige, and for thousands of years cows have also provided us with meat, milk and leather, not forgetting the butter, cheese, yogurt and tasty ice cream that we can now also make from milk.

Modern cows – all 1.4 billion of them in the world, according to the UN – are efficient meat- and milk-producing animals. The average cow produces more than 20 litres of milk a day. As children, we learn that cows can do this because of the endless hours they spend feeding on grass or hay, and it's likely that the very first pictures we saw of cows were of them happily grazing in green fields. And, where there are cows, there are cowpats. Cows are great at digesting plant material and producing milk, but this means they're also good at producing a lot of methane – a greenhouse gas implicated in global warming – and cow dung. Yes, this chapter is all about poo!

Cows belong to a group of plant-eating mammals known as ruminants, which includes giraffes, deer, antelope, sheep, goats and, of course, cattle. All these animals can acquire nutrients from plant-based foods because they can digest cellulose, the tough fibrous material that makes up the walls of plant cells. It's something that sets the ruminants apart, for most other mammals, including humans, don't have this ability.

The key to how it works lies in the design of a cow's stomach. After we humans chew and swallow our food, our stomachs serve as a holding tank, where digestion begins. Here, the food starts separating into simple carbohydrates, amino acids, small fat globules, vitamins and minerals. As the food passes out of the stomach and into the small intestine, this breakdown continues, and the body starts absorbing these nutrients. This basic process is also true of cows, but they have a few extra steps along the way. Most importantly, whereas humans have just one compartment to their stomach, cows have an impressive four.

The cows you see grazing in green fields are swallowing their food without too much chewing. The food then travels into the cow's first stomach chamber, known as the rumen or 'paunch.' Imagine a massive bag full of grass or hay, and this will give you a good picture of what this first chamber looks like. Inside, the partially chewed food forms grassy balls, known as cud, which are brought back up into the cow's mouth and chewed with a bit more

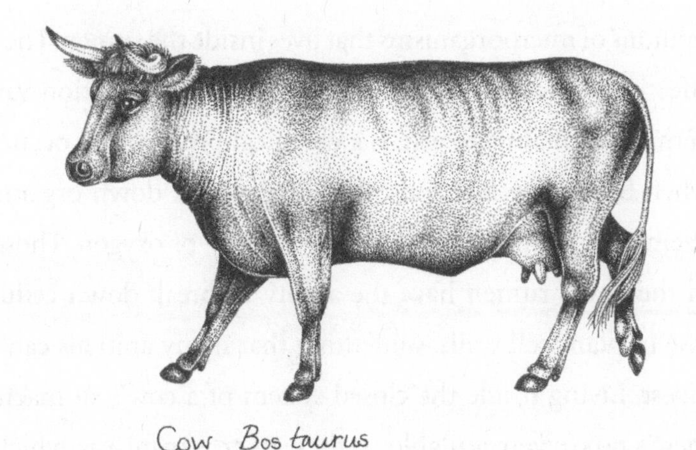

Cow *Bos taurus*

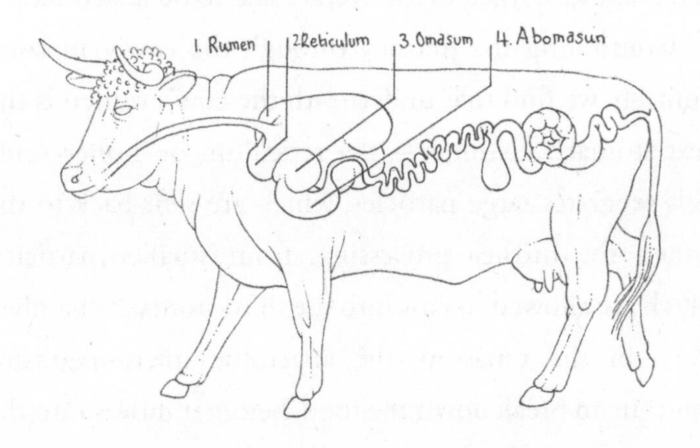

vigour. You might be familiar with the expression 'chewing the cud,' without realising it actually refers to this process of regurgitating and re-chewing grass, so the food is ready for the next stages of digestion.

At this point, our cow is able to absorb some nutrients directly from the rumen, but the real work of digestion is actually performed not by the cow itself, but by a team of millions of microorganisms that lives inside the rumen. They effectively turn the stomach into a giant fermentation vat. Fermentation, by the way, is a chemical process that occurs when bacteria, yeasts and other fungi break down organic chemicals, such as glucose, in the absence of oxygen. Those in the cow's rumen have the ability to break down cellulose in plant cell walls, something that many animals can't digest. Living inside the closed system of a cow's stomach, there's no oxygen available, so these microorganisms, which don't require oxygen to survive, are said to be 'anaerobic.'

Continuing the journey through the cow's amazing stomach, we find that underneath the cow's rumen is the next stomach chamber – the reticulum or 'honeycomb.' This separates large particles, which are sent back to the rumen for further processing, from smaller particles, which are allowed to pass into the third stomach chamber. Here, in the omasum, the anaerobic microorganisms continue to break down the food, before it moves into the final chamber, the abomasum. This last chamber acts more like a human stomach, digesting what's left of the food, which then passes through the small and large intestine, where nutrients are absorbed, before the waste is egested as cow dung.

This ability of cows to digest grass has captured the interest of a team of engineers and scientists in India, because the inner workings of a cow's stomach could help us dispose of human waste in a far more efficient and ecologically friendly way. The team lives and works in the southern Indian city of Bangalore, where something unpleasant has been going on.

Nestled on a plateau 900 metres above sea level, the modern Bangalore is probably most famous for being home to India's thriving Silicon Valley. It's also known as the city of lakes, because of its complex irrigation and drinking water system which was built by its rulers in the sixteenth century. Over the past few years, though, Bangalore's residents have been subjected to a troubling and seemingly impossible sight: these once beautiful lakes have been bursting into flames, sometimes for 30 or more hours at a time, the ash raining down on houses up to 10 kilometres away, and it's happening with unfailing regularity. To understand how water can catch fire, we need to understand what's in these lakes.

Once clean enough from which to drink, the lakes are now full of poisonous industrial and solid domestic waste. As Bangalore has grown in size, the lakes have been used increasingly as a dumping ground for raw human sewage, which in turn has encouraged the overgrowth of vegetation and weeds, with the production of methane from their oxygen-starved waters. Altogether, this has made the lakes highly flammable, which has been of particular concern to local IT engineer Tharun Kumar, who saw the health of his family declining as they breathed in these harmful fumes. Each day, nearly 400 million litres of raw sewage was being pumped into his local lake, simply because the city's sewage treatment plants were either broken or were considered too expensive to run. So Kumar decided to gather together a group of scientists and engineers to find a solution to the problem that would be both cheap and effective.

The team – who called themselves ECOSTP – noticed that most traditional sewage treatment plants work by using oxygen-dependent bacteria to breakdown the waste. This means that air needs to be circulated constantly through the treatment tanks, using fans and pumps, but this requires electricity, and electricity costs money. So, naturally they looked to nature for a solution, and they turned their attention to the cow, and, in particular, the inner workings of the cow's stomach.

The ECOSTP team noted that it was anaerobic bacteria that were breaking down food as it passed through the different chambers of the cow's stomach. Inspired by this, they decided the way forward was to develop a method of treating sewage that would mimic this process. To appreciate their challenge, we need to take a journey through the various chambers of their new sewage plant.

Firstly, the sewage is collected in a large tank, which is sealed shut, and anaerobic bacteria from cow dung is added. This is known as the 'settling chamber'. It's just like the cow's first stomach chamber, where the microorganisms get to work, breaking down the waste. Once the solid part of the sewage has sunk to the bottom of the tank, the rest of the waste flows down a pipe into the second chamber, where it mixes with more anaerobic bacteria. As it relies on gravity to flow down a pipe, no motors, fans or pumps are needed, which means there's no need for electricity. By the time it leaves this second chamber, the wastewater is now much cleaner, although, if you pour it into a glass, it looks like dirty water. At this point, it's still not safe enough to be released into local waterways, so it flows down another pipe into a third chamber, where it passes through gravel and bacteria for a further clean. The final stage is a gravel filter planted with local grasses and algae, which remove any remaining solids, bacteria, viruses and nutrients. The end result is a glass of clear, clean water.

There's also another plus. Not only does this cow-inspired solution mean that the sewage treatment plant doesn't need power, it can also operate without staff, bringing down costs even more. But it does have one problem: all those different chambers and pipes need double the amount of space as a normal treatment plant, and this is a bit of an issue, as space is at a premium in a crowded city like Bangalore. The team, however, had a solution.

They developed a way to hide the main sewage chambers beneath buildings and roads, and put the final plant and gravel filter stage of the process in gardens and green open spaces. Using this method, they've already built 20 of these new sewage treatment plants in new housing developments throughout the city, and they have plans to build even more. It comes at a time when there's a great demand for more innovative sewage treatment plants like this. In India, it's estimated that 93 per cent of human waste is dumped directly into rivers, lakes and the sea. Digging a little deeper into the subject, we find that across the world, according to the UN, a staggering 80 per cent of wastewater is discharged into the environment untreated. Since Kumar's team developed the new treatment plant, it's been officially recognised by the UN, and the team is now exporting its idea to other countries, encouraging them to build their own plants using local materials.

Across the world, unsafe drinking water and poor sanitation are a significant cause of diseases and death, particularly in children, so it's truly amazing to think that a cow eating grass in a green field has been the inspiration for this cheap, ecologically friendly, and sustainable solution to cleaning dirty water.

21

Pollution Solution: Manta Rays

I was sitting on the edge of the dive boat, my fins gently swirling in the water, looking out across the vast deep ocean, its intense azure blue matched only by the sky above. I focused my thoughts – as any free-diver would – on controlling my breathing. The calm conditions made this day absolutely perfect for a once-in-a-lifetime encounter. My exact location – often hailed as the little brother or sister to the Great Barrier Reef – was the Ningaloo Reef, in northwest Australia, which is located in some of the most food-rich waters in the world, and attracts a number of migrating marine animals. Humpback whales and whale sharks pass through, as well as the magnificent animal for which I was waiting, said by some to resemble a devil, and others an angel. The sound of a breathy gasp suddenly broke the silence. It was one of the support divers, surfacing. It snapped me out of my daydream, and back into reality. He called out: 'They're right beneath us!' It was time to dive.

I slipped gently into the water. Making even the smallest of splashes would've startled the beautiful creatures I was soon to meet. Floating at the surface, I took a few deep breaths, before plunging head first into the murky blue. I peered through the gloom, hoping, waiting, and then, as if by magic, there they were: three massive manta rays.

Everything slowed down, as if in a dream. A sense of calm came over me. It was the way the manta rays were circling in the water, swimming next to each other in tight formation, like spaceships from an alien world. Viewed from above, the dark upper surfaces of their bodies meant they almost disappeared from view as they blended in with the darkness of ocean; only then, betrayed by their lighter white bellies, flashing white as they rolled and banked through the water, could I see them once more.

Their flattened bodies, like living wings, were perfectly adapted to glide through water with minimal resistance. It was like watching an underwater ballet but, I came to realise, they weren't circling in such an ethereal way to amuse me. They were in the middle of a feeding frenzy, although, with nothing else in sight, where or what was their prey? The clue lay in the murky water. The nutrient-rich waters of Ningaloo are filled with tiny microscopic plankton, and it's this that gives the water its hazy appearance, and also draws in the mantas. They were sifting plankton out of the water, in a process known as filter feeding, and you can't help but notice another of their impressive features: their mouth, which is a huge letterbox. If you didn't know any better, you'd be forgiven for thinking they might swallow you whole... and that was how I got to peer inside the mouth of a manta ray. Now, there's a sentence you don't hear yourself say every day!

Giant oceanic manta rays really are giants. With a 'wingspan' up to 8.5 metres, they are the largest rays in the world. The name *manta* means 'blanket' or 'cloak' in Spanish, and describes the appearance of the animals' flat, diamond-shaped body and triangular wing-like pectoral fins. Mantas also have two smaller cephalic fins, which stick out at the front of their head. These can curl up to look like horns, and this has given rise to their other name: 'devil fish.'

For many years, scientists thought there was just one species of manta ray but, in 2008, they discovered there were actually two distinct species, and now it's been suggested there may be a third. There's the reef manta ray, which is generally found along coastlines of the Indo-Pacific region, as well as off the coast of tropical East Africa. The giant oceanic manta ray occurs in all the world's oceans, and spends much of its time out at sea, far from land. And, finally, there's the Caribbean manta, although we're still learning about this potentially new species.

What we do know is that all manta rays belong to a much bigger group of animals called rays, which number more than 500 species, and include sawfishes, skates, electric rays – oh, and did I mention, they're also related to sharks. Rays differ from sharks in their flattened disc-like bodies, and their five gill openings and mouth are generally found on the underside of the body. Many rays – but not mantas in this case – swim and breathe differently from sharks, propelling themselves with their pectoral fins and taking in oxygen for breathing through large openings called spiracles on the upper surface of the head. The ray's tail is also generally long and slender, and many species, including the Caribbean manta ray, come armed with one or more sharp saw-edged venomous spines that can be used to inflict painful wounds.

As filter feeders, manta rays move through the water by the rippling rise and fall movement of their pectoral fins, which drive water backwards and the manta forwards. Much like some species of sharks, known as obligate ram ventilators, they have to swim continuously in order to keep oxygenated water passing over their gills. When it comes to feeding, seawater enters their large, forward-facing rectangular mouths, where plankton are captured, and water escapes through the gill openings. Exactly how manta rays filter plankton out of the water has captured the interest of a team of scientists at California State University in Fullerton, California, and the Whitney Laboratory for Marine Bioscience at the University of Florida. The first question that marine biologist Dr Misty Paig-Tran asked was, 'Why don't manta rays need to clear their throats?'

If the manta's filters worked liked sieves, there ought to be some evidence of clogging, similar to what happens when you drain your pasta in a colander: smaller bits of food often get stuck in the holes. What the researchers discovered is that manta rays use a previously unknown method of filtration, which causes the particles to bounce over the filter system, instead of going through it. This means it doesn't need to 'clear its throat', or its filters in this case, because they rarely get clogged. Let's take a look inside.

As the ray swims forwards, water is pushed into its mouth, and lining it are tiny, angled slats called gill pads. When seawater rushes over them, it forms mini-whirlpools between each pair of slats. These whirlpools or vortices don't suck the particles down, instead they push them up, preventing the pieces of plankton and other particles from falling into the crevices. As a result, the particles bounce off the slats, and become more and more concentrated in the throat, whilst the water drains away. The ray then swallows this concentrated soup of plankton, so the particles never actually go through a filter – which has got me thinking, maybe we should call them bounce-feeders instead?

Back on dry land, the team ran a series of particle experiments in order to visualise the bouncing. They washed coloured dye over plastic versions of the manta's gill-pad structures. They also designed a number of mathematical models to confirm and better analyse what was going on.

In case you're wondering, there *are* animals, such sardines and mackerel, which do seem to simply sieve the water to capture their food, but what's far more exciting for the scientists is a filtration system that doesn't clog up, like that of the manta ray, because this technique could be used as a model to collect microplastics, and prevent them from polluting our oceans.

Plastic pollution is one of the most serious environmental problems we face as a species. According to the Clean Seas Campaign of the United Nations Environment Program, about 13 million tonnes of plastic enter our oceans each year. That's the equivalent of dumping the entire contents of a rubbish truck into the ocean every single minute. In 2010, scientists predicted that the amount of plastic entering the ocean could increase tenfold by 2025 and, even now, it's possible that every creature in the sea might come into contact with microplastics at least once over the course of their lifetime. As we're yet to fully understand the long-term effects of swallowing microplastics, the consequences, if not reversed, could be disastrous.

Dr Paig-Tran and her students ran some studies in their lab and found that, provided food was available, some animals actually preferred the microplastics to the uncontaminated plankton food source. They wouldn't, though, if it weren't for chemical cues on their surface. Large pieces of plastic also take years to break down into microplastics, the entire breakdown process releasing greenhouse gases, which contribute to climate change. Scientists have also raised concerns that chemicals in plastics, as well as chemicals which attach themselves to plastics in the natural environment, could lead to poisoning, infertility and genetic disruption of marine life. It has the potential to affect humans too, if we were to unknowingly start ingesting large quantities through the seafood we eat.

Microplastics are undoubtedly a nightmare, because they easily get into our wastewater. The microfibres from washing synthetic clothing, such as fleeces and other sports kit, are present in the wastewater from washing machines, from where they eventually make their way to water-treatment plants. The ability for a treatment plant to deal with these particles depends on what kind of microplastic it is.

There are three types: some are negatively buoyant, so they sink to the bottom of treatment tanks, and are passively treated as they don't travel beyond that tank. Others are positively buoyant, so they rise up to the top, where they can be scooped off the surface. That leaves us with the third type – the microplastics that are neutrally buoyant. These are the ones that float about in the middle of the water, roughly the same level that zooplankton occurs, and it's these microplastics that aren't treated or removed, because water systems aren't equipped to deal with such neutrally buoyant particles. Imagine, though, if we could invent a clog-resistant filter, modelled on the manta ray's mouth, which could be installed in treatment plants to catch these remaining fragments of plastic. Then, we'd finally have a solution for removing microplastics from the natural environment.

The manta ray filtration system has also captured the interest of a team of students at The Hague University of Applied Sciences in the Netherlands, who were looking at a filter system to collect microplastics in rivers, which would be one way to prevent the pollution reaching our oceans.

The team discovered that there are just 10 rivers around the world that together contribute to a staggering 90 per cent of the plastics that enter our oceans. They also found that these same rivers are a big source of all three types of microplastic particles – negatively buoyant, positively buoyant and neutrally buoyant. These rivers also have two things in common. Firstly, there's a large population of people living in the surrounding area, hundreds of millions in some cases. Secondly, it's a situation that's made worse because of substandard waste-management systems in those places. If we could better target these rivers, and somehow extract the plastic, then the team hope to reduce the amount of contaminants reaching the ocean. What they came up with is a filtration system, called the 'Floating Coconet,' which aims to do just that.

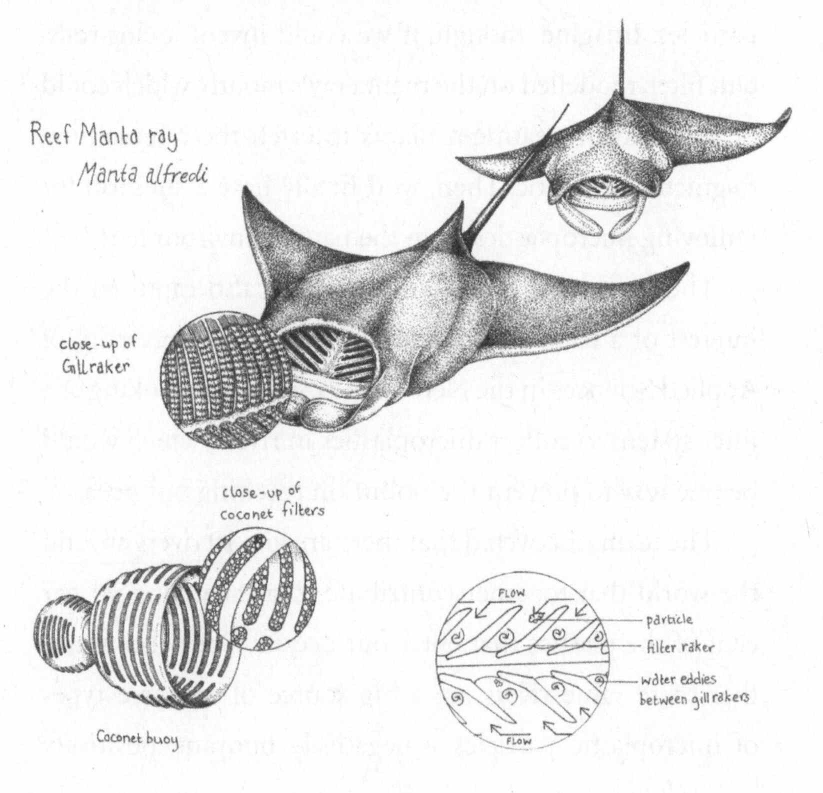

Their design is based on a series of basket-like funnels that float in the water column. Each basket is nearly two metres long and, looking at it from the side, it consists of a larger container at the front, with a smaller container behind, to which a net can be attached. The opening at the front, which is about 1.5 metres in diameter, guides the river water and any plastic contaminants inside the basket. Its inner walls have a gill-like structure covered with fins, and these create tiny currents, similar to those whirlpool eddies in the manta ray's mouth. The device aims to direct and collect free-floating microplastics into the net.

What's interesting here is that the fins inside can change position. By pulling on a lever, they can go from a forward-facing position, in which they collect plastic, to a backward-facing position, where they'd release the captured pieces. This provides the added benefit of being able to release any small pieces of plastic that do get stuck between the fins, simply by pulling the lever and flicking the fins back and forth to dislodge any blockages. With this strategy of capturing swirling particles, while flushing out the water, the team hope to gather plastic particles inside the basket, which would then flow into the net. When the net's full, the waste can be removed, and recycled.

By linking several floating baskets together, the team have suggested that multiple 'Coconet Trees' could be held in place, using ropes attached to heavy concrete blocks on the riverbed. The idea would be to create a dynamic moving wall across a river, a bit like a seagrass meadow. The easy swaying motion of the floating baskets means that fish and other marine life could easily swim around them. The design is still very much in its infancy, but the team have really been inspired to continue to innovate and experiment with other ideas, regardless of whether or not their filtration model goes into production.

Meanwhile, Dr Paig-Tran and one of her colleagues, James Strother, have a US patent pending for a type of ricochet filter, the technology of which was based on the filters of the manta ray. If they're successful, applications would include a mechanism to remove microplastics from wastewater, as well as microplastic removal from beer- and wine-making, which to be honest, I didn't even realise was an issue. There's even a collaboration on the cards, with the Alden Research Laboratory in Colorado, to create fish egg exclusion screens, which could be used to screen water entering cooling water intakes for power stations, for example. It's all very clever stuff, and all this just from peering inside a manta ray's mouth!

22

Glues from Life to Save Life… and Make Cupboards!

When I was young, one of my all-time favourite things to do was to go on trips to the seaside with my grandma. She'd sit down in the sand with all the other grandmas, armed with her cooler full of drinks and food, and I'd go splashing in the water with my brother and sister and all the other kids. Building sandcastles right by the edge of the water was the best. As the waves came closer, we'd dig out the biggest moat to try to stop the castle from crumbling and, while waiting for the tide to come in, I'd go from rock pool to rock pool, meticulously investigating what treasures lay beneath the curtains of seaweed. Fast-moving crabs were always a big hit, but after a while I began to notice other creatures. What first appeared to be a small white rock, was in fact a slow-moving limpet. I'd peer in closer at the underwater garden using a stick to gently swirl the water. All of a sudden, a coral-coloured starfish would begin to limber up. Sea anemones, like stringy jelly sweets that had come to life, wafted food towards their open mouths. I've spent hours searching for the perfect seashell, and tried, in vain, to nudge barnacles off the rocks. I even considered collecting some mussels for dinner, although, I left them right where they were in the end. Firstly, because I knew my grandma wouldn't have been best pleased with me carrying a soaking bag of smelly mussels on the coach ride home. And, secondly, I had trouble prising them off the rocks, like Kaichang Li, who lived a half a world away, in Newport, on the Oregon coast.

Kaichang Li and his friend had been looking for crabs along the rocky shore, but they had no luck; then, as he was wading in the surf, he spotted a bunch of blue mussels anchored to the rocks, clinging tight against the turbulent waves. Determined not to leave the beach empty-handed, the two friends sat down amongst the rocks and instead tried to collect mussels. Li, who is a chemist at Oregon State University, was amazed at how much effort was needed to pull the shellfish off the rocks; in fact, the men failed to pull away any by hand. They had to use a large piece of driftwood to prise some of them free. Whilst his friend took his share of the mussels home to eat, Li took

Blue mussels *Mytilus edulis*

his portion back to the lab. How, he wondered, did they manage to grip the slippery rocks?

The blue or common mussel is a bivalve mollusc, which means it has two shells hinged on one side. Individuals can be coloured purple or brown and, of course, blue. They're to be found on rocky shores in the North Atlantic and North Pacific, where they attach to rocks, as well as to piers and docks in sheltered harbours and estuaries. They may look alike from the outside, but there are separate male and female mussels. You can tell them apart because, when you open them, males generally have pale gonads containing sperm, while female gonads have the brighter orange colour of eggs. The colour difference is not 100 per cent accurate. Mussels living on the upper shore can be darker orange, no matter the sex, the colour linked to the greater environmental stress experienced by individuals exposed to the air for a large part of the day.

Mussels are described as semi-sessile animals, as they can move. They are able to detach and reattach in order to reposition themselves. Dense masses attach themselves to surfaces by what looks like a beard. This is made up of 50 to 100 individual filaments called byssus threads, which are produced by a byssal gland inside the shell. The threads are not only very tough, but they can also heal themselves when damaged. The surface or cuticle is as hard as the epoxy resin used in the manufacture of printed circuit boards, but it still has the ability to flex and extend. At the end of each thread is a small adhesive plaque, and it is with this that the mussel clings to the surface of rocks.

Back at his laboratory in Oregon, Li discovered that the threads were made from strings of proteins, which were themselves made from amino acids whose composition is unique. It was a light bulb moment – could he use this knowledge to make super-sticky glue? A good adhesive must be strong, of course, but also flexible, and mussels had evolved a perfect solution: unlike any other type of adhesive, it worked well in water.

Now, whilst Li was pondering his mussels, Steve Pung, vice-president of technology at Columbia Forest Products, was thinking about plywood. Hardwood plywood is made from several thin sheets of wood that are glued together. The result is a versatile and highly workable material with strength and flexibility, and it's used in all sorts of ways in the home, from floors to cupboard sides. The problem is the plywood manufacturing process involves the use of formaldehyde, and there's growing evidence of a link between formaldehyde and cancer. Steve and his associates were keen to find something that could replace it.

Li, meanwhile, was eating tofu for lunch at the university, and it got him thinking about soybeans. Soy or soya, he recalled, is about 50 per cent protein, so could he convert the soya protein into mussel protein to create an effective adhesive? He compared the chemical structures of soy protein and mussel protein, and invented a new type of glue by mixing soy flour and a curing agent in water. A curing agent is used to harden a material, and the agent he used was the one that gives paper towels and tissue papers greater wet strength.

Li then made plywood that was very water resistant, and, to demonstrate this, he boiled samples of it and then dried them in the oven. They didn't fall apart. Wood swells when wet and shrinks when dry, so this swelling and shrinking would have put incredible pressures and stresses on the glue. It not only remained sticky, however, but also waterproof, just like the mussels that are soaked by the sea and dried by the sun. 'The adhesive penetrates the wood fibre,' Li explains, 'and locks it in place. It's like you put a key in a lock and turn it, and then can't get it out.'

When happy with the result, Li gave a presentation about his work, and guess who was in the audience – Steve Pung. Steve couldn't quite believe his ears. Here seemed to be the answer to his problem, an answer that he'd been looking for, for years. Given that there's no time like the present, Steve immediately got talking to Li after his presentation, and by the time he left they'd struck up a deal to pursue commercialisation of Li's adhesive. At first, the process didn't work. Li might have given up there and then but, like the mussels he'd been studying, once attached to an idea, he was not going to let go.

Eventually a solution was found, and Pung's company started converting their mills. Pressure on wood mills to avoid formaldehyde means that other companies are adopting the soy-based resin-bonding process. It makes for a healthier working environment and formaldehyde-free products. Just think, if Li and his friend had caught crabs that day on the beach at Newport, he might never have stumbled across those mussels, nor have imagined where his curiosity about their tenacious grip would lead him.

SAVING LIVES

Any kind of surgery on the human body comes with an associated level of risk, but the decision whether or not to perform surgery on an unborn baby in the womb is undoubtedly one of *the* most challenging in medicine. According to Diederik Balkenende, a member of the Messersmith Research Group at the University of California, Berkeley, one of the greatest risks in performing surgery on a foetus isn't the surgical procedure itself, but getting into and out of the very fragile amniotic sac, the fluid-filled bag which surrounds and protects the foetus. Sealing this sac after surgery is no simple task, but the humble mussel might provide scientists and surgeons with an answer.

A few years ago, surgeons would've had to cut open the abdomen and uterus in order to perform such procedures. Today, doctors can use endoscope tools to carry out an entire operation through a small hole. Even so, to reach the foetus, the instruments have to penetrate the protective amniotic sac. These incisions are larger and pose a greater risk than membrane punctures created by amniocentesis. During surgery, the membranes become exceptionally vulnerable at the point where they're punctured, and what makes things even more complicated is that the fluid-filled membrane is difficult to stitch back together, and doesn't heal. It's like trying to sew up a water balloon. But what if we could use a type of glue to prevent the amniotic sac from tearing? Could this help the foetus stay in the womb for longer, and lead to a healthier future for the baby?

Sealing the membrane with a conventional adhesive or surgical glue poses its own challenges: the membrane of the amniotic sac is wet, and delivering a surgical glue through a small hole at the end of the operation is in itself very difficult. To make things even trickier, the developing foetus is biologically sensitive, so you definitely don't want to use any harsh chemicals, or anything that could affect the growth or health of the baby. To try to solve the problem of wet adhesion, researchers have turned again to the mussel for inspiration. The insides of the human body, and the amniotic sac where the foetus develops, are watery environments, not so different from the ocean, where the mussel routinely and successfully glues itself in place on rocks.

Phillip Messersmith and his colleagues studied the proteins in the byssal threads in order to design their own synthetic-mussel-glue. What appears to play a key role in the mussel adhesives is a substance with the tongue-twisting name L-3,4-*dihydroxyphenylalanine* or L-DOPA for short. It's a naturally occurring amino acid, not only a precursor to neurotransmitters such as dopamine, but also important in producing sticky adhesive proteins. Messersmith's team have included it in their synthesised adhesives, and they're working to increase their strength, as well as the effectiveness of adhering to wet surfaces. They added a few more compounds, which the mussel itself uses to make glue, including other amino acids and metal ions (iron and silver), and hope that will further improve the glue's performance.

To test their glue, one of the researchers, Sally Winkler, used pieces of the membrane that surrounds a cow's heart as a model of the amniotic sac. They applied their synthesised glue with a syringe to overlapping pieces of the wet, filmy tissues taken from a dead animal. When it made contact with moisture on the tissues, the mixture immediately became rubbery, and it only took about an hour for the glue to set and hold the pieces together. The researchers have also carried out tests on the thin membrane lining the inside egg shells in order to see how well they stick under wet conditions, how fast they harden, and importantly whether they'll hold in place. But, even with the right polymer and solvent, the researchers still needed to figure out how their glue solution might work in a real surgery.

It's clear that repairing a hole in an amniotic sac is a daunting engineering challenge, so Messersmith's group are approaching the problem of how to deliver their glue from a new angle. They describe it as 'pre-sealing,' and it's really quite clever. First, a needle would be used to make a tented space between the wall of the uterus and the foetal membrane without puncturing the membrane. Then the glue is injected into this space where it's left to set into its rubbery state. The injected sealant starts to harden, or cure, and attach both to the foetal membrane and to the uterine wall. The surgeons could then penetrate the amniotic sac with their surgical instruments through the area covered by the sealant patch. The team believe this should minimise and prevent the spread of any damage to the membrane during surgery. This membrane reinforcing also allows a watertight seal to close around the surgical instruments.

Routine clinical use of the glue is some way off, as the researchers are still perfecting their solution but, if it's successful, pre-sealing could be used for other delicate surgeries, for example on the bladder, the spinal cord sheath, intestines or other places where surgeons need to puncture through delicate tissues or prevent leaks.

GECKO'S FOOT ADHESIVE

Another route to adhesives that could be used inside the human body comes from studies of a small lizard – the gecko – and its amazing ability to walk up vertical surfaces, even those as smooth as glass, and scurry across ceilings. Their secret lies in the billions of hair-like structures, known as setae, which line the undersides of their toes. Imagine a tree, and how its trunk splits into branches, and these divide and divide again into ever more branches; something similar occurs on the underside of the gecko's foot. The toes have a series of ridges, which are covered by the setae. Each one is finer than a human hair and divided into hundreds of 'split ends,' called 'spatulae.' In this way, the surface area of

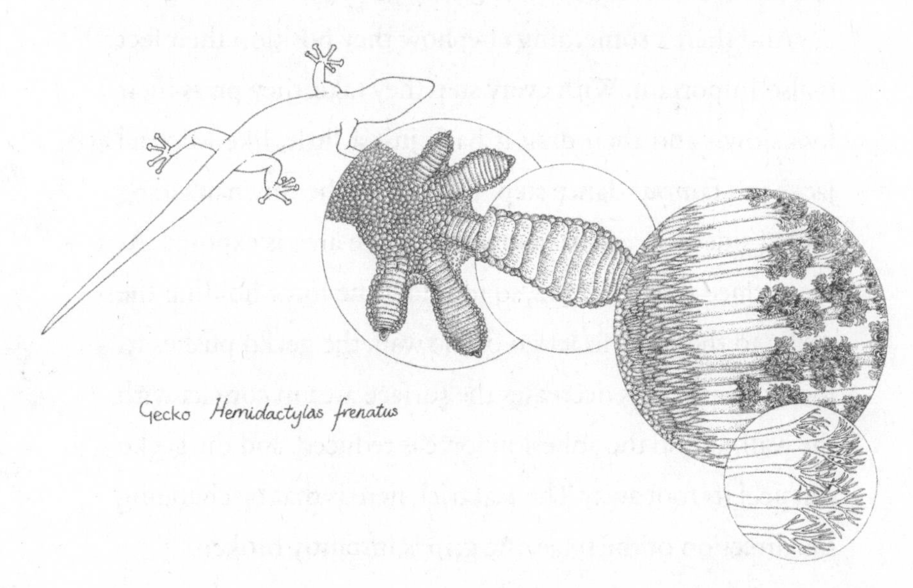

contact is maximised, the actual sticking being down to a phenomenon known as Van der Waals force.

Johannes Diderik van der Waal was a Dutch physicist who published his thesis in 1873. Put simply, he proposed that the positive sides of one molecule attract the negatives sides of another, with the result that the molecules are drawn together, a bit like a couple of magnets. The molecules on the setae in the gecko's foot and those on the surface on which it's moving are charged and attracted to one another, and this results in this adhesive, gravity-defying effect. Each square millimetre of a gecko's foot has about 14,000 setae, so the adhesion forces are so strong that a gecko could effectively walk upside down, carrying a rucksack, weighing 40 kilograms...

And there's something else: how they position their feet is also important. With every step they take, they press their foot down and then drag it back just a little, like Michael Jackson's famous dance step. This causes the tiny hairs to be pulled sideways, and so a greater surface area is exposed. As the surface area increases, so too does the force holding the gecko to the wall. To let go of the wall, the gecko pushes its foot forward. This decreases the surface area in contact with the wall, and so the adhesion force is reduced, and the gecko can peel its foot away. The real trick here is that by changing the direction of the setae, the grip is instantly broken.

As you can imagine, this has attracted a lot of interest amongst scientists and, in the last ten years or so, scientists have been able to create synthetic setae. Dr Jeff Karp is a bio-engineer who together with colleagues at the Massachusetts Institute of Technology, Boston, has created a waterproof adhesive bandage inspired by geckos, which could soon be used in hospitals to join sutures and staples for patching up surgical wounds or internal injuries. When their work began, very few adhesives were used internally, but there was a need for some sort of 'tape' which you could wrap around the intestine for example after an operation, or that could be used to patch up a hole. It would also need to be biodegradable, breaking down into substances that weren't toxic, and it would need to be flexible and compatible with the body, so that it wouldn't cause any internal inflammation. This led to the development of a tape with tiny micron-sized hair-like structures, each less than a millimetre in length. The hairs spaced just wide enough to grip and interlock with underlying tissue, and then a thin coating of glue was added to help the tape stick in wet places, like the heart or bladder. There's also potential for infusing this gecko-inspired tape technology with drugs, which would be released as the tape degraded inside the body.

KEEPING STICKINESS

And, before we leave stickiness, there's another creature helping to solve the problem of glue in damp environments – the spider. The dilemma this time is finding a sticking plaster that really is 'waterproof'. The reason that sticking plasters seem to lose their stickiness and peel away from the skin when you swim or get them wet in the bath or shower is down to 'interfacial water'. This is water that gets between the glue and the surface to which it's supposed to be sticking. It forms a slippery and non-adhesive layer. This interferes with the formation of adhesive bonds between the two layers, the glue and the surface, so overcoming the effects of interfacial water is one of the challenges facing designers of commercial adhesives. At the University of Akron in Ohio, USA, researchers are looking to spider silk for an answer.

The silk threads of spider webs are coated in a type of glue – a hydrogel, which means it's full of water. You might think this wouldn't be sticky, but it's quite the opposite; in fact, it's one of the most effective glues in the natural world, even when things get hot and humid. The research team began by examining glue produced by orb web spiders, and discovered it is made of three elements: two specialised proteins called glycoproteins;

a collection of LMMCs, which stands for 'low molecular mass compounds,' which don't weigh much; and water. The glycoproteins act as important binding agents to the surface, and glycoprotein-based glues are quite widespread in nature: in adhesives associated with fungi, algae, diatoms, starfish, sticklebacks and English ivy.

The chemicals that are really interesting, though, are the LMMCs. They do the clever stuff. The LMMCs absorb water. They attract or absorb moisture from the air, which keeps the glue soft and tacky and perfect for sticking to things. Perhaps more importantly, the LMMCs also move water away from a boundary, so that the glue can stick to a surface, even in high humidity. The ability of spider glue to overcome the problem of water by effectively absorbing it is key, and it's this chemical property that could have commercial potential, including the manufacture of a sticking plaster that sticks come rain or shine.

23

Cats and Road Safety

It's late at night. You're walking alone through a deserted town. Above you, the streetlamps cast a ghostly glow. It starts to rain hard. You flick up your hood and pick up the pace, as you make a turn down a dark alley. Here, away from the streetlamps, you struggle to see, but remember the torch in your pocket. You turn it on, and begin to navigate through the gloom. That's when you hear it – a distinct rustling right behind you. You quicken your step, but the sound seems to follow, keeping up with your every pace. Your heart beats faster and faster, your senses alert. With nowhere left to run, you decide to confront the horror stalking you. Turning quickly, you raise your torch, and that's when you see it: two bright penetrating spots of light staring through the dark directly back at you. Phew! It's only a cat. I'm sure – well, maybe not as dramatically as I've just described – that you're familiar with the way a cat's eyes reflect light back to you in the dark, and, I'm pretty confident you're familiar with this, because, let's face it – humans love cats.

The ancestors of domestic cats are believed to have evolved in Asia, migrating to Africa, Europe and America around 11 million years ago, but it was about 10,000 years ago, when humans began cultivating crops, that cats moved in much closer to us to take advantage of the mice and other rodents that were attracted to our stores of grain. This meant that cats – true to their independent and intelligent-minded stereotype – chose to domesticate

themselves, and so began our long-running relationship. We humans are so drawn to cats that some cultures regard them with the highest esteem, even as gods. In ancient Egypt, for example, the cat goddess Bastet was believed to bring fertility, prosperity and happiness. In Norse mythology, the goddess Freya is said to have travelled around in a sled pulled by two giant white cats. In Myanmar, cats are still seen as the sacred guardians of Buddhist temples. In Japan, they're believed to possess the souls of our dead ancestors.

If you've ever lived with a cat, all of this might sound a bit overenergetic. You're probably more familiar with just how much they love to nap. The average cat sleeps between 12 and 16 hours a day, curled up on the sofa, under your feet, or purring away on your lap. Yes, cats sleep anywhere and everywhere, but far from being lazy, they need to do this because, when the sun goes down, cats come alive.

Cats are nocturnal animals – they hunt at night. Although, they're also known for being crepuscular, which means they can be most active during the low light of dawn and dusk. To be effective at this time, they need really good eyesight. As such, a cat's eye has adapted and evolved to be far better at seeing at this time of day, than, say, our human eye. It's also the reason that they glow so brightly, in a spooky kind of way, when you shine a light in their eyes. It was this particular quality that was noticed and copied to create a small, but highly effective invention that has kept millions of road users safe all over the world. This is the story of how cats' eyes inspired, well… cat's eyes.

In this book, most of the stories of animals inspiring new technologies are recent, some still in the development stage, but the use of cat's eyes in road safety has its origins about 100 years ago. The invention is so commonplace now that we hardly think about it, but it's had a massive impact on our wellbeing, so let's take a closer look at the inspiration behind the invention.

The eyes of a cat are really quite beautiful. There's something beguiling, even mysterious, about the way a cat will stare at you for what seem like an age, and then slowly blink. This is apparently its way of showing that it likes you, although, it's hard to tell sometimes with cats. Up close, it's easy to see why we find cat's eyes so attractive. If you look into your own eyes, you'll see the pupil at the centre – the dark, circular opening that regulates how much light is let in. Beyond this is the retina, the light-sensitive layer that lines the back of the eye. It's composed of special cells, called photoreceptors, which convert light into electrical nerve impulses that are sent via the optic nerve to the brain. Here, they're interpreted as images, and, that's how we see the world around us.

On a bright day, when there's lots of light, our pupils contract, because there's plenty of light for our photoreceptors to work well; in the dark, they dilate to let in as much light as possible, so we can see more clearly. A cat's pupils do the same thing. The most striking thing about cats' eyes, though, is how their pupils, rather than being round like ours, are more like vertical slits. It's what gives a cat that elegant and distinguished look. Compared to our pupils, these vertical pupils react much faster to light and change shape. They can also expand and become much bigger to allow in more light, which, of course, is very useful when you're out chasing mice at midnight. Cats have, however, one more awesome adaptation, and it's this that gives their eyes that wondrous, almost luminous glow, when you shine a torch into them.

When light enters the cat's eyes, some of it will hit the photoreceptors of the retina directly, and cause electrical impulses to be sent to the brain – just like our eyes – but the cat's eyes have an extra property. Some of this light travels through the retina, and hits a very special layer of tissue behind it. This is called the *tapetum lucidum*, Latin for 'bright tapestry.' As the name suggests, it is a highly reflective surface. Any light that hits it is reflected back through the retina to be picked up by the photoreceptor cells. In this way, the *tapetum lucidum* increases the amount of light landing on the retina, which helps the cat to be an exceptional hunter in the low light of dusk and dawn. Some of the reflected light passes back through the retina, but instead of being absorbed by the photoreceptors, it exits the cat's eyes, and we see it as eye shine – the eyes appear to glow in the dark.

Cats aren't the only animals with a *tapetum lucidum.* You may have noticed the same glowing effect in dogs, horses and birds and even in the eyes of little spiders hiding in the grass; but we humans don't have a *tapetum lucidum*, so our eyes don't shine in the dark in the same way. It did, however, inspire a clever, life-saving invention, and to understand how it came about, we need to go back to 1930s Britain.

Can you imagine what the roads were like in the early 1930s? Cars had only recently overtaken the horse and carriage as the main mode of transport, and road safety was most definitely a work in progress. There were street-lights in towns and cities, but in the countryside there was only the moon to guide you through the twisting, turning lanes, and that, of course, was not a reliable light source. A man in Yorkshire in the north of England, however, was about to change all that.

According to one account, mechanic and inventor Percy Shaw found himself driving down a particularly dangerous road, a section of what is now the A647 called Queensbury Road, on his way home to Halifax one foggy night. It was a steep hill, with a sharp drop to one side, and the only thing guiding and, indeed, stopping him from driving over the edge was a small fence. Shaw was more accustomed to driving at night in towns, where his headlights would be reflected in the metal tramlines embedded in the roads. This helped him find his way more easily; on this country road, there was only darkness. At this point in his journey, Percy Shaw tells of a startled cat, which jumped onto the fence ahead of him. Shaw's headlights were immediately reflected back by the *tapetum lucidum* in the cat's eyes. Shaw stopped his car to investigate, and realised he'd been driving on the wrong side of the road and had been very close to falling over the edge. If it wasn't for the cat, he could have had a very nasty accident.

Shaw had already considered how one might move the reflective studs on road signs to the road surface but, with the cat encounter, he realised he'd stumbled across a groundbreaking idea. What if he could have a whole row of cats sitting in the road, with their ever-present reflective eyes guiding drivers? He spent the next few years developing and refining an invention that would do just this. By 1934, he was ready to patent his new road safety device, which he named appropriately 'cat's eyes.' He set up the Boothtown-based company Reflecting Roadstones Ltd to manufacture the device.

To build his cat's eyes, Percy Shaw used a new type of retro-reflecting lens. This is a special lens that reflects light directly back to its source, and the reflected light is much brighter than that coming back from a diffuse reflector. They had only been in use for a few years, when they helped to light up advertising billboards at night. It works because of the way it's made. Like an ordinary mirror, the front is made from glass. The difference is what's behind. An ordinary mirror will have a flat-mirrored backing, but the retro-reflecting lens has a backing composed of a curved mirror. It can also have two or more flat mirrors set at different angles to each other, which means it can reflect light back from a wide range of angles.

Shaw's original design had a pair of these retro-reflecting lenses set into a protective white rubber dome, set in a cast-iron base. When light shone into it, it mimicked the *tapetum lucidum* of a cat and reflected light directly back, in this case, to the driver of an oncoming car. They were placed at regular intervals along the middle of the road, where the white line divides the traffic coming from opposite directions. The idea was that if a car was driving at night, its headlights would be reflected by the cat's eyes, and this would help the driver keep on course. The rubber casing was flexible enough to give if a car accidentally drove over it, preventing it from being broken.

A later design incorporated a small reservoir to collect rain-water, which helped to keep the lens clean. The casing also provided both audible and physical feedback for drivers who wandered too close to the middle of the road.

At first, Shaw found it surprisingly difficult to persuade the authorities to invest in his invention, but then came the Second World War. To avoid becoming the target of aerial bombers, Britons were told to switch off all lights at night, although cars were still allowed to drive, albeit with reduced headlights. Suddenly, Shaw's cat's eyes were in great demand, as they were the only things helping drivers to navigate through dark city streets and country roads. After the war, they received firm backing from the Ministry of Transport, and the British government ordered that cat's eyes be installed throughout the country. Soon, Shaw's invention was in use not only in the UK, but also around the world. Different countries built on the design, with some using different-coloured lenses to mark out the difference between the centre line of the road and the kerbside, or to indicate exits and entrances to slip roads. Although the designs have changed over the years, one thing remains constant: where cat's eyes have been used, millions of lives have been saved. I think you'd agree: it's just the sort of cool, clever type of invention you'd expect from a cat.

24

High-Rise Sponges

Here's a question for you. What do orange elephant ear, branching tube, brain, stinker, branching vase, chicken liver, red boring, brown encrusted and lavender rope all have in common? The names of Harry Potter spells? No, not quite, although that would be a pretty good guess. They're all, in fact, names for a fascinating group of organisms, known as sponges, which can all be found in the Flower Garden Banks National Marine Sanctuary in the Gulf of Mexico, along with the touch-me-not, variable boring sponge, loggerhead and orange lumpy encrusting! Yeah, I know it does sound like something from a J.K. Rowling novel, but trust me on this one. I kid you not!

Sponges are one of the simplest animals on the planet and they've been around for a very long time. They certainly date back to the Ediacaran period, as fossils have been discovered in rocks 580 million years old. These ethereal, delicate-looking formations had early naturalists convinced that they were plants. It wasn't until 1795, when their feeding method was described, that they were finally recognised as animals. So, what kind of animal are they?

Sponges belong to the phylum Porifera, which comes from the Latin *porus* meaning 'pore' and *ferre* which means 'to bear'; so we have 'to bear holes', which, once you set your eyes on the elaborate structures they create, makes perfect sense. They are thought to be the first group of animals to have branched off from LUCA – the Last Universal Common Ancestor, which gave rise to all of life, which makes the sponges a sister taxon to all the other plants and animals that ever lived on Earth.

When it comes to different species, a review in 2012 indicated 8,553 known species, and they come in all sizes and in every colour imaginable – bright purples, blues, yellows and reds. The smallest are just a few centimetres long, while the very largest is a gargantuan deep-sea sponge, resembling a folded blanket, which is 3.7 by 2 metres, and was discovered in the Papahānaumokuākea Marine National Monument off Hawaii in 2016. Sponges can also be unbelievably old. In the Antarctic there are giant volcano sponges, about two metres tall, which are thought to be about 15,000 years old, making them the oldest living things on the planet.

Sponges are ubiquitous. They're found all across the world, from the shallows to the deep sea, where they occur in virtually every type of marine environment, and some are found in freshwater too. Even though they're classified as animals, sponges are unlike most other animals. They have no digestive tract, no sensory region, or even any true tissues; in fact, they're made up of no more than a cluster of cells. What keeps them together is a 'skeleton' made of a flexible network of fibres, together with rigid elements called spicules. The spicules are made of a glassy type of silicon or from calcium carbonate, the same material as chalk, and they can be organised to form incredibly complex structures. To understand them, we can, however, consider something a little simpler.

The next time you're in the kitchen or in the bath, grab hold of a sponge and feel how soft and flexible it is. This soft texture is very much like the fibres found in a living sponge. These days, most household sponges are made from synthetic materials, but they resemble the natural sponges that were once commonly collected for human use. The small holes you see on the surface of a natural sponge are called ostia, and these holes are where water enters the living sponge. The larger openings on the surface are called oscular, and these are where filtered water and waste are expelled out of the sponge.

You might think that sponges just hang around doing nothing all day – and, with a name like red boring, you're not going to inspire a sense of excitement – but you'd be wise to think again. It's possible that they were the first ever ecosystem engineers, which means that they have the ability to change the environment around them. They're great, for instance, at filtering ocean water. They have the ability to clean the water close by and help cycle nutrients of which other organisms can take advantage. In nutrient-poor coral reefs, sponges even help increase the availability of carbon, an important element needed by many organisms to grow. The sponge discharges a form of 'sponge poop', which other organisms feed on, and which, in turn, promotes the health and biodiversity of the reef.

Even more extraordinary is that sponges are the only animals that, if broken down into tiny groups of cells, have the ability to reassemble themselves. The cells can find one another, clump together and build another sponge. As if all of that wasn't enough, it turns out that sponges might be the perfect source of medicines. The cells of sponges, together with the tiny living animals inside them, can produce complex chemical compounds that may have the potential to treat pain and inflammation, as well as diseases like cancer and Alzheimer's.

For reproduction, sponges have two options. The first is called asexual reproduction. This is when a piece of the sponge separates from the main organism and grows into a new sponge, which is essentially a copy of the original – a clone. Alternatively, there's sexual reproduction, which involves sperm fertilising eggs. While adult sponges are generally stuck in one place, their larvae can disperse in the sea and swim to new locations.

Sponges aren't so boring after all, then. What's more, their skeletal systems have enabled different species to live on hard rocky surfaces and soft sediments like sand and mud. It's these complex structures that have attracted the interest of a team from the Harvard John A. Paulson School of Engineering and Applied Sciences, as they could inspire the next generation of what could be the tallest skyscrapers and longest bridges ever to be built.

The research team, led by Professor of Applied Mechanics Katia Bertoldi, chose to study the skeleton of a deep-water marine sponge called *Euplectella aspergillum*, because its structure was very resistant to 'buckling,' an easy-to-understand term used in engineering to describe when a structure becomes deformed in some way, which results in it bending and twisting out of shape. The skeleton of *Euplectella* is a stunningly beautiful tube-like structure made of silica. The walls, rather than being solid,

Venus's Flower Basket *Euplectella aspergillum*

are patterned with holes, like fine lacework. It's one of a group of sponges known as the glass sponges, and its English common name is Venus' flower basket, from its horn-like structure, which is commonly associated with Venus, the Roman goddess of love, beauty and prosperity.

The sponge is found in the deep sea around the Philippine Islands, with similar species occurring near Japan and other parts of the Western Pacific. The animal's skeleton or basket is a curved tube, about 25 centimetres long, and is covered in a tuft of hair-like spicules at the narrow base, which anchors the sponge to the soft sediments of the sea floor. The skeleton of Venus' flower basket is highly prized by collectors. In Japan, for example, it's regarded, as a symbol of commitment, due to the fact that each basket often has hidden inside a mated pair of shrimps, which spend their entire lives trapped in there. It's a gift often given to newlyweds. The research team, however, was interested in this sponge because of the robustness of its skeleton. Trust the scientists to ruin a perfectly good love story, but, to be fair to them, they weren't looking for love; they were looking for inspiration.

If you've ever driven across a bridge or tried to assemble a set of metal storage shelves, chances are you'll be familiar with a diagonally reinforced lattice structure. A basic lattice is a structure in which strips of wood or metal crisscross each other, leaving square or diamond-shaped gaps. You'll often see this used for screens or fence panels to support growing plants. The extra diagonal strips in a reinforced lattice are a simple and cost-effective way to keep the basic structure stable. American architect and civil engineer Ithiel Town originally patented the structure in the early 1800s. He wanted a way to build sturdy bridges out of lightweight and cheaper materials. As one of the Massachusetts-based team says: 'It gets the job done, but it's not optimal, leading to wasted or redundant material, and a limit on how tall we can build.'

They wondered, however, whether they could create lattice structures that were more structurally efficient: in other words, use less material to achieve the same strength. Could Venus' flower basket provide them with the answer? After all, these sponges have had more than half a billion years to perfect their skeletal systems.

Taking a closer look at the glass sponge, the engineers could see that, in order to support its tubular skeleton, *Euplectella aspergillum* has an extra layer. It's like when you want to keep a bulky load together, and you wrap some rope tightly around it. This sponge has the equivalent of not just one rope, but two sets of parallel rope-like rods or struts that wrap around it. These struts overlap one another, and are joined to the layer underneath. To picture it, imagine a mat of squares all made of fine wire. Now roll the mat into a tube. Then imagine pairs of fine parallel wires wrapped around the tube, starting from the top and working down to the bottom. Finally, have another group of paired wires wrapped in the same way, but in the opposite direction, so that the wires crisscross over each other. This is pretty much a simplified version of how the sponge looks.

Even though previous studies have examined the behaviour of the individual hair-like elements in preventing cracks from spreading and in resisting buckling, there wasn't much known about the benefits of this extra double layer. So, in a series of computer simulations, the team copied this design and compared the mechanical properties of the sponge skeleton to other existing lattice structures. The sponge design outperformed them all, and was able to withstand far heavier loads before buckling. What I found really cool is how the Venus' flower basket creates this lattice.

When the living sponge grows, it goes through two phases: a flexible phase and a rigid phase. In the early flexible phase of growth, the vertical, horizontal and diagonal struts remain separate, and aren't fused to one another. This gives the tube-like structure the ability to expand. Once the maximum length and width of the lattice tube is reached, however, the sponge begins to deposit a glassy cement which fuses all the pieces of the skeleton together, creating the final rigid form.

From their studies, the team concluded that it's the 'ropes' – the diagonal reinforcements – of the sponge that are responsible for the resilience of the structure. They demonstrated that these reinforcements improved the overall strength by more than 20 per cent, without the need of additional material to achieve it. It's another perfect example of how Nature has evolved a better lattice design than any of our existing patterns, and this knowledge is going to come in really handy, when it comes to constructing more effectively, taller buildings and longer bridges, and also in aerospace engineering where lighter, stronger structures are as valuable as gold dust. Amazing to think that the designers of the next skyscraper or award-winning bridge might have Venus' flower basket to thank for its sturdy, lightweight design.

25

Camels and Cool Medicines

With its tall slim legs, its long neck that dips downwards and rises up again to meet its small head, and, of course, its famous humped back, it's hard to mistake a camel for anything else. There are three different species. The most common is the one-humped Arabian camel or dromedary. It accounts for about 90 per cent of the world's camel population, and was domesticated 4,000 years ago; even today most are owned by somebody. The only 'wild' dromedaries are feral animals that were introduced into Australia in the mid-nineteenth century, escaped their masters, and made their home in the Outback. Breeding unchecked, they have become a serious environmental problem, with a huge impact on native vegetation, and so they are culled when things really get out of hand.

The second species is the domestic Bactrian camel, which was domesticated even earlier, probably between

5,000 and 6,000 years ago. It is a two-humped camel native to Central Asia, where its tolerance of cold, drought and high altitudes made it the transport of choice along the ancient network of trading routes known as the Silk Road, which was travelled from the second century BCE until the eighteenth century. These long journeys gave rise to both the dromedary and the domestic Bactrian being called 'ships of the desert.'

Bactrian Camel
Camelus bactrianus

Dromedary
Camelus dromedarius

The third type of camel, the wild Bactrian, also with two humps, is the only truly wild species but is critically endangered. It survives – but only just – in the Gobi Desert, where there are thought to be fewer than 1,000 individuals. Counting them is difficult, because they are notoriously shy. They run away swiftly from anything that has a hint of humans, and keep on going without looking back. Their range stretches across Mongolia and northwest China, and, unlike their cousins, they've never been domesticated. What I find particularly fascinating about them is that some will happily survive drinking water with a higher concentration of salt than that found in seawater, which their domestic cousins cannot. Not only are they mystifying animals but, with so few, they are one of the most endangered large mammals on the planet.

The dromedaries and the Bactrian camels are in the biological family *Camelidae* – the camelids, which also includes the guanacos, vicuñas, llamas and alpacas from South America; in fact, the camelids have their origin in North America about 45 million years ago, and there they remained until about 2–3 million years ago, when a land bridge joined the two continents and the camels' ancestors headed over to Asia and onward to Africa, while the guanacos and their cousins took advantage of the bridge that formed where Panama is today and became part of the Great American Interchange.

Here's a cool camel fact: modern-day camels evolved from creatures that inhabited the Arctic, including a giant camel that lived 3.5 million years ago in the Canadian Yukon. It was about 30 per cent bigger than modern camels, and wore a very shaggy coat. Today, many of the adaptations to extreme weather probably came come from these Arctic ancestors, although the region at that time was a little warmer. So why are there no camelids in North America today? The simple answer is nobody knows, but it's thought that either they died out because people hunted them to extinction or their demise was the result of climatic changes at the end of the last Ice Age.

While the very earliest camel ancestors started out as relatively small animals – one was the size of a rabbit, another not much bigger than a goat – today's camels are all big. At nearly two metres at the shoulder, a fully grown dromedary stands taller than a horse, and the domestic Bactrian is a touch taller. How do you remember which is which? Here's a neat trick: if you think of the letter D lying on its side, it's looks a single hump, whereas the letter B lying on its side looks like a double hump, so, D stands for dromedary and one hump, with B standing for Bactrian, which has two! The question to which I've always wanted to know the answer, though, and I'm sure you do as well, is what exactly is inside the hump? To answer that, we need to do a spot of myth busting.

It's often said that these humps are full of water, which is what allows camels to survive lengthy periods in the desert, but this is completely false. The hump is full not of water but of fat, and to understand what's going on we need to put on our lab coats and see what's happening at a molecular level.

Fats are made up of molecules called hydrocarbons, which, as their name suggests, consist mainly of hydrogen and carbon atoms. They contain large amounts of energy, and the camel's body – much like ours – will break down, or metabolise, body fat in order to release this energy. The cool thing about this process – of which camels take full advantage – are the by-products of this metabolism: the hydrogen and carbon atoms combine with oxygen to release energy, carbon dioxide, and – wait for it – water.

What's even more ingenious about this process is the fact that for each gram of fat that's broken down, just over a gram of water is released. So, we can see that this is an extremely efficient way of storing that all-important H_2O. Camels, though, have a problem retaining all of this water. The oxygen needed to break down the fat also requires the camel to increase its breathing rate; overall, it actually ends up losing some water in the form of water vapour, which escapes into the surrounding atmosphere via their lungs.

Nevertheless, a camel's hump enables it to effectively store food and, in a roundabout way, water. How long the animal can survive on its stored fat depends on how active it is and on the weather. The size of the hump can change depending on how much food the camel eats. When food is scarce, the camel's body uses up the stored fat, causing the hump to lean over and droop. There's also the suggestion that storing fat on their backs, rather than in a fatty layer across the entire body, is a clever way to avoid overheating, with the added benefit that the hump offers some protection and shade from direct sunlight.

With adaptations like these, camels are able to survive in some of the most hostile and extreme environments on the planet, where finding water is only one of many challenges. Finding sufficient food is not that easy either. Camels feed on a variety of grasses, as well as green shoots and plant stems and twigs. Their mouth has a hard top palate and the cheeks and tongue are lined with protective papillae so that tough, thorny desert plants can be ground down, like using a mortar and pestle. Even all this becomes a minor challenge when you think about the extreme temperatures that these animals must endure. To help them cope, Bactrian camels, like their Arctic ancestors, grow a great shaggy coat in the winter to protect themselves against the freezing cold, and then shed it during the hot summer.

In severe heat, a camel can go an entire week – sometimes more – without water; and can survive even longer without drinking if it can obtain some of the moisture it needs from the plants it eats or if it's not working so hard. It can survive a four per cent weight loss, which is about 19 kilograms for an adult Bactrian, but it can also take in large quantities of water in a relatively short time. A thirsty animal can drink 130 litres in just 13 minutes – that's a whole litre every six seconds, but please don't try this at home!

Another adaptation is their ability to minimise sweating. You might think that having a fur coat is the last thing an animal needs in the heat, but if a camel lost its thick fur, it would utilise about 50 per cent more water. The fur provides great insulation, handy on cold desert nights, but also important when things get hot. If the external temperature is so much greater than the camel's internal body temperature, this blanket prevents heat from moving inwards. With their fur coat acting more as a barrier to external heat, it's only in the hottest weather – when its body temperature has risen to 41°C – that a camel needs to sweat. This is the perfect evolutionary adaptation because, not only does it reduce heat gain, but it also reduces water loss.

In many ways, the camel's secret to a successful life in the desert lies in its ability to store a lot of heat before it needs to sweat. During the day, the body temperature of a fully hydrated camel can range from 36°C to 38°C, while at night it can fall to 34°C. This drop at night gives the camel a buffer, so it can 'store' more heat the following day, before it needs to sweat at 41°C. With this strategy, it might only need to sweat for a couple of hours at the end of the day, which means the camel can achieve a further saving in water. In the hot, dry desert, a camel – which can weigh at least five times as much as a human being – only uses a quarter of a litre of water each hour.

This ability to keep cool has attracted a lot of interest from scientists who are looking at new forms of insulation, and the timing couldn't be better. According to the International Energy Agency, the global energy demand for cooling – which is measured by things like the demand for fridges, freezers and air conditioners – is projected to triple by 2050. As a result, there's increasing interest in the different ways we can keep things cool without using energy – so called 'passive cooling.' Given that more than 10 per cent of the world's population has no access to electricity, passive cooling provides us with a fantastic method of distributing and storing things like food and temperature-sensitive drugs. One of the most promising solutions is based on evaporation from hydrogels, a kind of gel or jelly that absorbs and retains large quantities of water. They can also release water through the process of evaporation. Evaporation is what happens when you get hot and sweat, with water turning into water vapour. It's the evaporation of your sweat that cools your body down, so you could say that hydrogel mimics the sweat glands of animals. This is a prime example of passive cooling.

Scientists have been interested in hydrogels for some time, because the water they absorb can be released through evaporation and, more importantly, without the need for an external power source. The problem, though, is that the effect is short-lived, so the challenge is how to make it last for long periods of time. A research team led by Jeffrey Grossman at the Massachusetts Institute of Technology, USA, has turned to camels for its inspiration. They've combined a hydrogel layer with a thin layer of another gel, called an aerogel – a light, porous insulating material – and this evaporation-insulation double layer mimics the biological cooling system found in camels.

The hydrogel layer on the bottom is like the camel's sweat gland. It allows water to evaporate, providing a cooling effect. The aerogel layer above plays the same role as the camel's fur, providing crucial insulation. It prevents heat in the environment from passing through, but, because it's porous, it still allows water from the hydrogel to escape. By achieving evaporation and insulation at the same time, the cooling period is significantly extended, and the two gel layers – or the bilayer, as it's called – is only one centimetre thick.

The team tested their double-layered gel in the lab, using a special temperature and humidity-controlled chamber. They discovered that the bilayer was able to cool an object to a temperature 7°C lower than its surroundings, and, when compared with a single hydrogel layer, the length of time this bilayer could keep an object cool increased by 400 per cent. As Jeffrey Grossman points out, this translates to over 250 hours or 10 days of cooling. The next step is to make the material more scalable. In other words, they need to find a way of more easily producing it in larger sizes and quantities. The team has suggested their design could also help cool buildings, reducing their total energy consumption.

When you consider how often we need to transport food and medicines – not only within a country, but also globally, especially when there might be little or no access to fridges or a power source – it becomes clear how valuable a passive cooling system like this would be. It would be positive news for the environment, and also would have the potential to become a life-saving technology – all this from a desert-dwelling champion.

CAMEL NOSE AIR CONDITIONING

There are, of course, people who traditionally share the desert with the specially adapted plants and animals, and they have also had to adapt to these harsh and arid conditions. Take the Bedouin peoples of the Arabian Desert, for example. Their nomadic lifestyle means they don't stay in one place long enough to exhaust the few resources that are there – good for the environment – and when they change locations, they take their tent homes with them. These are cleverly constructed to allow air to circulate within them, keeping them cool in the heat of the day, and warm when temperatures plunge at night.

In the modern world, humans have conquered the desert. Think of the gleaming, glittering desert cities of Las Vegas or Dubai – I've been to both and they're quite spectacular, stunning feats of human engineering, and hardly places of discomfort or hardship. It seems there's nothing you can't do in these awe-inspiring human-made environments; in fact, it's easy to forget you're even in a desert, especially when you're watching breathtaking performances at the casino theatre, or zipping down the slopes in an indoor ski resort – yes, you can go skiing on snow in the desert – mind-blowing!

All of this building, though, means a high consumption of fossil fuels, just to keep the hotels, casinos and ski-centres cool. But what if you could design a more environmentally friendly building? That's exactly what a team of architects in Egypt was wondering, and they turned to the camel for a solution, and more specifically to its nose. It turns out that the camel's nose is an ingeniously designed, biological, air-cooling device that's another key player in the camel's survival. As the saying goes, if you're in doubt, trust your instincts and follow your nose, and that's precisely what the architects did.

If you were to peer up a camel's nose, you'd come across a structure that's very different to the inside of our noses. Where we have one passageway in each nostril, a camel's nose has an intricate labyrinth of many channels, all covered in a moist mucous membrane. This enables the camel's nose to act as both a humidifier and dehumidifier within the same breathing cycle. In other words, as the camel breathes in and out – it can both add moisture to the air and take moisture from it. In this way, it helps keep itself cool, as well as hydrated.

When the camel breathes in, air travels into its nasal passageways where the moist membranes add water, cooling the air down. You'll be familiar with this effect if you've ever held up a wet cloth in front a fan. As the air blows through the cloth, it comes out slightly cooler on the other side. This is exactly what's happening to the air when the camel breathes in. This cooler air travels into the camel's lungs where it remains at body temperature. When the camel breathes out, this air is full of water vapour. Normally, in humans and other mammals, this vapour is breathed straight out into the atmosphere. It's one of the main ways we lose water from our bodies. Living in an environment where water is extremely scarce, however, the camel needs to hold on to as much of this water vapour as possible.

This is where those membranes in the nasal passageways come into use again. They're not only moist, but they're also covered in a special water-absorbing substance that's able to extract moisture away from the air; so, as the camel exhales, they start absorbing water vapour from the outgoing breath. As the camel has so many of these membrane-covered nasal passageways – in other words a lot of surface area – this increases the amount of evaporation and condensation that takes place when it breathes in and out. The result is that the camel retains nearly 70 per cent of the water that would normally be lost in a breathing cycle.

The process also helps to keep their brain cool. This is because the cool air in the nasal passageways lowers the temperature of the blood circulating around them, before it travels up to the brain. This leads to a significant level of cooling in their brain tissue, helping camels to thrive at temperatures that would be lethal to other animals.

The architects, led by Dr Merhan Shahda at Port Said University, wondered if they could copy all this. After several months of design and experiment, they came up with a special camel-nose-inspired cooling system, which could be incorporated into new buildings or fitted onto existing ones, and operate it without the use of costly electricity. The first part mimicked what happens when a camel breathes out and collects water vapour. The architects constructed a glass box, shaped like a triangular prism, which was placed on the southern side of a building, the side that – if you're in the northern hemisphere of the world – receives the most sun. The roof of the prism – the sloping bit – was connected by a hinge, so that it could be opened. Inside, the floor was covered in a layer of calcium chloride, a material that has the ability to absorb water from air.

Desert air contains more moisture in the cool of the night, than it does during the baking heat of the day, so the prism was designed such that its sloping roof would swing open at night, to let in this moist air. Once the air was circulating inside, it came into contact with the layer of water-absorbing chemical. This would act just like the mucous membranes inside the camel's nasal passageways when it breathed out. In other words, it would absorb water vapour from the air.

To get this water out of the calcium chloride, the prism relied on the natural heat of the desert. Remember, this prism had been placed on the south side of the building where it received the most sunlight. To increase this effect, the architects surrounded it with a curved reflecting panel. Then, as the sun shone during the day, the sloping roof was shut, and the reflective panels concentrated the sun's energy directly onto the layer of calcium chloride. This caused the water, which had been absorbed, to evaporate and rise towards the sloping roof. Once there, it cooled and condensed into droplets of liquid water, which ran down the inside of the slope towards a pipe at the bottom. This drained away, into a water-collecting tank below.

Once they'd worked out how to harvest water in this way – this is in the desert, remember – the architects set up a system to use this water to cool the building. For this bit, they mimicked what happens when camels breathe in. To do this, they used two natural materials: straw and burlap, a type of coarse hessian canvas. These were draped inside a ventilation hole, just below the glass prism and its water tank. Small holes were then made in the tank so that water would drip down onto the straw and burlap below, keeping it moist. Finally, a solar-powered fan was set up to draw hot air towards the straw and burlap from outside.

The idea was that, as the hot air travelled inside, the wetted straw and burlap would add moisture to it. This copied the way the camel's mucous membranes add moisture to the air when it breathes in. And, as this air picks up moisture, it naturally cools down. Because straw and burlap are made up of lots of fibres, they have a large surface area, like the camel's nose, which, in turn, increases the cooling effect. The result is a building in a hot desert that's kept cool, without the need to burn fossil fuels.

There was one final part to the system. The architects made special holes at the top of the building so, as the air inside the building began to heat up, the hot air would rise and escape naturally. This created a constant air flow of cool air coming in through the straw and burlap covered ventilation hole, on the side of the building, with warmer air leaving through these top floor vents.

This enterprise is still in development, but early tests show that it's been really successful, both in lowering the temperature inside buildings and in adding much-needed humidity to the air. The architects hope that, in time, their design can be deployed in other countries with deserts, pulling water out of thin air for free, and cooling homes and offices without the need of electricity. The hope is that this sort of technology will allow people to live more in harmony with nature, just as the Bedouin have done for centuries in their mobile ventilated tents. And, all this new development is because of the camel – not only the trusty ship of the desert, but also the inspiration behind a new harmonious way of living amongst the desert dunes.

26

Lobsters and Space Telescopes

What do deep space and deep water have in common? Even asking that question fills me to the brim with excitement, because I love anything and everything to do with space, from finding out the answers to age-old mysteries of the universe to what new worlds we might find that humanity might one day even call home. Space has been famously described as the final frontier, but the truth is the same could be said of the Earth's deep oceans. Both are the last great unknowns, full of secrets yet to be discovered.

If you think about, it's amazing that we humans, a species that shares 98.8 per cent of our DNA with chimpanzees, have the ability to send probes into outer space. Over 40 years ago, NASA launched two such probes – *Voyager 1* and *Voyager 2* – and they have long since passed by all the planets, moons, asteroids and comets of our solar system and headed into interstellar space, billions of kilometres away. They've travelled the furthest any human-made object has ever been. Yet, incredibly, they are still sending back information.

Likewise, the exploration of inner space – the deep ocean – is an astonishing adventure closer to home, yet still with all the logistical challenges that those exploring outer space must face. Manned and unmanned submersibles – basically fancy submarines that can withstand great pressures – can now reach the greatest depths, such as the Mariana Trench in the western part of the Pacific Ocean. At more than 11 kilometres below the surface, it's the deepest of all the deeps, where the pressure is about a thousand times the standard atmospheric pressure at sea level. More people have been on the moon than have reached its deepest part, the Challenger Deep, but, actually, wouldn't it be wonderful to go down there and, for that matter, also experience the weightlessness of space, say, at the International Space Station? Well, one person – British born explorer Roger Garriot – was wealthy enough to do just that, and threw in a visit to the North and South poles, just for good measure – the first person on Earth to do so.

On his deep dive, Roger would have peered out of the reinforced viewing port and seen all manner of deep-sea creatures down there, including shrimp-like crustaceans called amphipods, relatives of crabs and lobsters, that live at these crushing pressures. To bring it full circle, it is one type of crustacean in particular that has inspired a new way of exploring space. The animal in question can be found in many oceans, including in deep water, but, to be honest, you're probably more familiar seeing it on your dinner plate – perhaps deliciously drenched in butter or in a roll, with French fries on the side. You've guessed it! This chapter is all about lobsters and, let me assure you, there's far more to this fascinating creature than just a delicious plate of food.

I like to think of lobsters as the aristocrats of the sea, and that's not just because they have blue blood or, more accurately, blue haemolymph. It's blue, incidentally, because they use a copper-based molecule, haemocyanin, instead of the iron-based haemoglobin like we do, to transport oxygen around their body. They're also special because they've been swimming in the seas for a very long time, around 360 million years. Down the eons, many different types have evolved, but it's the large-clawed lobsters, in the genus *Homarus* that evolved about 100 million years ago, that we tend to associate with the word 'lobster.' There are two types living today – the American and the European. We think of them as being bright, pink-red, but that's the colour they turn when they're cooked. In its natural habitat, the American clawed lobster is more of a brownish colour, while the European species is generally a shade of blue.

Lobsters, like all crustaceans, are invertebrates. They don't have a backbone or internal skeleton, but instead have a hard protective exterior 'shell' or exoskeleton, to which muscles are attached, just like our internal skeleton. They have five pairs of walking legs, each covered in tiny hairs, which they use, along with their antennae, to sense their environment, and it's the lobster's front pair of legs that develop into its distinctive large claws.

It may surprise you to learn that lobsters can be right clawed or left clawed. It sounds like something totally made up, but it's true, and all depends on which claw a lobster prefers to use. The dominant claw is bigger, thicker and is also known as the crusher claw. It's used for dispatching prey – fish, mussels, clams, worms and other crustaceans – and for battling other lobsters. The smaller claw is the cutter, and used to hold or slice up food.

If all this sounds a bit brutal, let me also tell you, that clawed lobsters have a more communicative side. They can signal to each other using chemicals secreted in their urine. To you and me, it might not seem the best way to talk, but for many animals – dogs, lions, bears included – this is all business as usual. For lobsters, this helps them to work out which lobster is the most dominant and also which mate to choose.

European Lobster
Homarus gammarus

Another interesting fact is that they live for a very long time. It's believed that some lobsters reach 100 years old. There's also the myth that the clawed lobster is immortal, because it never stops growing. Throughout its life it goes through a series of moults. This is the process in which it sheds its old exoskeleton, grows a bit, and then replaces it with a larger one that's been growing underneath. Although they do keep doing this throughout their lives, the moulting does slow down as a lobster ages, and eventually they die. Their long lives are still quite remarkable, though, because they remain strong and healthy well into old age, which, unlike us humans, doesn't seem to affect their ability to continue mating and having young.

For most of their life, they search for food and avoid being eaten, and, at certain times of the year – especially in spring and summer – visibility is limited because the water is clouded with phytoplankton. At other times, it might be silt and sediment, so they've evolved a particularly ingenious way of visualising their world. It doesn't provide them with sharp images, but it does help them to spot something worth eating or what they would perceive as danger. It's this adaptation that's been copied to help us see further into outer space.

The human eye operates by using a property of optical physics called refraction. Energy waves, such as light or sound, bend as they travel through one substance and into another. When light travels in through the curved, clear front part of our eyes, which includes the cornea and the lens, it bends or refracts in such a way that it becomes focused on a special sensory layer at the back of our eyeball – the retina. It's a thin layer of tissue covered in cells which converts light into neural signals that travel via the optic nerve to our brain, which we then interpret as the visualisation of the world around us. This principle works really well for us, because we tend to live in environments where there's plenty of available light. The lobster, however, spends much of its time in the gloom, so its eyes need to work a little differently. They've evolved to use reflection, instead of refraction.

Picture a lobster's eyes. Think of a pair of black, shiny, globe-like spheres set on movable stalks on either side of the lobster's head, a bit like a giant shrimp or prawn. Now, if you look very closely – and by that, I mean if you look through a microscope – you'll see that each of these eyeballs is covered in thousands of tiny, precisely arranged squares. These are the ends of more than 10,000, long tubes, with a square cross-section, that form a geometric pattern on the curved surface of the eyeball. The tubes have straight sides, which are flat and highly reflective, just like a mirror, and, just like thousands of tiny mirrors, they work by reflecting incoming light down to the lobster's retina. These mirrors are arranged all over the eyeball's curved surface, so they're

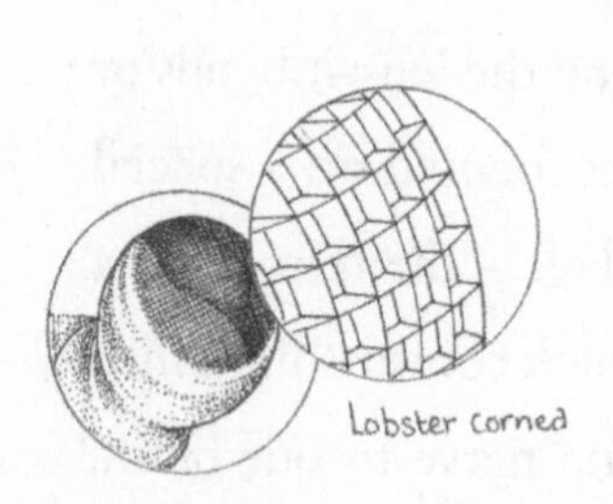

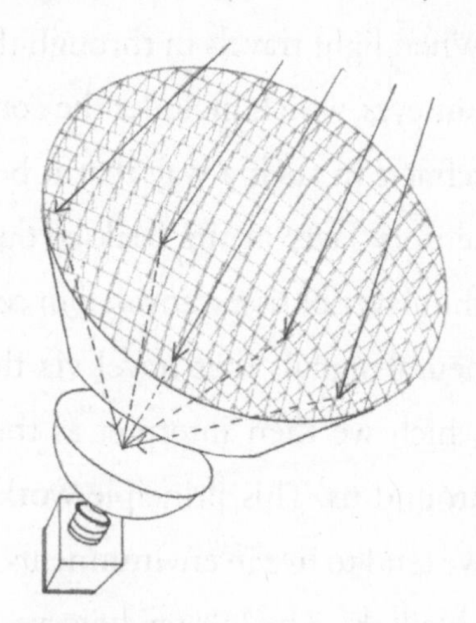

able to reflect light from a very wide angle, giving the lobster a 'field of vision' of 180 degrees. To put this into perspective, the field of vision for each one of our eyes is 150 degrees. This reflective arrangement is extremely useful, helping lobsters move around in an environment where visibility would otherwise be very poor. And it turns out that the inner workings of a lobster's eye are also pretty handy when it comes to monitoring events in deep space.

When astronomers look into space, one of the things they're hoping to detect is the presence of X-rays. But aren't X-rays the things doctors use to check if we have broken bones, I hear you say? Correct. An X-ray machine is able to see inside our bodies, because it utilises a type of electromagnetic radiation called X-radiation or X-rays, a penetrating beam of high energy that was discovered back in 1895 by German physicist Professor Wilhelm Röntgen. You can think of X-rays as being very similar to a beam of light but, in this case, they have a far higher energy level, which means, unlike light, these rays can pass through objects. This is exactly what's happening when we have a medical X-ray. X-ray beams are fired through our bodies on one side while, on the other, an X-ray-sensitive film picks up an image. Because materials like bones and teeth are relatively dense, they absorb more X-rays than skin and other soft tissue. Silhouettes of the bones are left on the X-ray film, while pretty much everything else – which is less dense – appears transparent.

So, what does all of this have to do with expanding on our understanding of deep space? Well, deep space is full of X-rays. High-energy objects, such as stars and black holes, emit them. The drawback for astronomers, though, is that these X-rays aren't detected on Earth, because our atmosphere is so thick that it absorbs them before they reach the ground. There is, however, a way around this: it is possible to detect X-rays using telescopes installed on satellites that orbit our planet and, crucially, can operate outside of our atmosphere.

A traditional X-ray telescope is a heavy piece of equipment, which can only look at a small patch of sky at any one time. This means you have to know exactly where the X-rays are coming from, in order to point the telescope directly at them. This is fine if you're measuring something you already know is happening, but what if you want to monitor a larger part of the sky for signs of unexpected activity? The answer is to build a telescope that works like the eye of a lobster. The idea was first conceived in 1977 by Roger Angel of the University of Arizona, and was the basis of an X-ray-based 'all-sky monitor.' Since then, it's taken almost 30 years of building on this concept by several universities around the world to perfect the optics technology.

So how does it work? Inside the 'lobster-eye telescope' is a thin, curved slab of glass that is covered in tiny tubes, just like the square tubes in a lobster's eye. X-rays enter these tubes and are reflected to a single point to create an image. Because the tubes are arranged on a curve – think of it as the top half of a sphere – they're able to pick up X-rays from a very wide angle. This mean a telescope like this is fantastic at picking up unexpected and fleeting flashes of X-rays from all over the sky. What's more, this type of telescope is much lighter than the traditional telescopes, which, when you consider that it has to be packed onto a rocket and sent into space, is a huge bonus.

There are now a number of lobster-eye telescope projects in development, in China, the USA and France and in the UK at the University of Leicester, working in collaboration with other international space programmes. These projects aim to observe a wide area of the cosmic sky, on the lookout also for another type of high-energy radiation called gamma ray bursts.

These are extremely powerful events that are believed to occur when a black hole is created, or when two stars crash into one another, or perhaps when a huge star collapses in on itself. Imagine the power! These are some of the most energetic things that can happen in the universe as we know it and, if you haven't already guessed it, these events also give off X-rays. The thought process is to use the lobster-eye telescopes to identify the general location of these gamma ray bursts. With this information, we can then direct another spacecraft or satellite, equipped with similar detectors, to look at that star or cosmic event in much greater detail.

What scientists are trying to understand is what happens the instant a black hole is created. This is important because it tells us a lot about the conditions around the star – how hot the material is, what the matter's doing, and how quickly it's changing. By surveying the sky in such a deep and thorough way, the lobster-eye telescope is helping to revolutionise the way we understand the universe. It's even – and this is very cool – helping to prove a theory developed by the most famous scientist of them all, Albert Einstein.

Back in 1916, Einstein predicted that vast intergalactic explosions would cause ripples to spread throughout the universe, much in the same way that dropping a pebble in a pond causes ripples to fan out in the water. These ripples are known as gravitational waves, and have only recently been detected. It's hoped the lobster-eye telescope will help future astronomers discover where they come from and the conditions in which they're formed. I think you'll agree, it's all a bit mind-blowing, and amazing to think that the observation and continued understanding of deep space is based on the optical powers of a lobster under the sea.

27

Waterloo Station and Pangolins

More than half the world's population live in urban areas, so it's more than likely that, when you're out and about, you'll be immersed in a metropolis landscape of large buildings and bright city lights. I think nothing illustrates our ability to master our surroundings more than the grandeur of the architectural marvels that dominate our city skylines. The idea that we can dream up complex structures in our minds and then turn them into something real and tangible is just awesome. Increasingly, as we build, we've looked to nature for inspiration. Humans have been copying geometry, patterns and principles of the natural world ever since we started putting roofs over our heads, and we've even built structures to actually mimic nature.

Take the Corinthian column, for example. This structure is the most elaborate of all the Greek columns: a cylindrical pillar with the capital decorated in leaf and flower ornamentations. These Corinthian columns, which adorn many Greek temples, have also been used to add a sense of grandeur to more modern buildings, such as the US Supreme Court and the US Capitol. The inspiration behind them is said to be the leaves of the *Acanthus* plant, also known as Bear's Breeches. Its story is rather charming, albeit tragic, and appears in the world's first books on architecture, *The Ten Books on Architecture*, published in the first century BCE. In the ten volumes, the Roman architect Vitruvius tells of a maiden of marrying age who lived in the city-state of Corinth and died prematurely. She was buried with a basket of her favourite things placed on top of her tomb, near the root of an *Acanthus*. The following spring, shoots, leaves and flowers grew up through the basket, and they caught the eye of the Greek architect and sculptor Callimachus. He included their designs on column capitals, and the columns that bear them became known as Corinthian columns.

There's also the Lotus Temple of New Delhi, whose white, marble-covered walls are designed to look exactly like the petals of the sacred lotus flower, and the famous Sagrada Família cathedral in the Spanish city of Barcelona, with huge concrete columns resembling the trunks and branches of trees. The cathedral's architect, Antoni Gaudi, believed that nature is the best designer, and, as well as copying Mother Nature, other architects have sought her guidance for solutions to newer more conceptual design challenges. Sponges, for instance, are helping with the design of lighter, taller, stronger structures, as you can read in Chapter 24, but in this chapter we're going to find out how one rather unusual mammal helped an architectural team in London squeeze a very complicated building into an extremely tight space. The animal is the pangolin.

Pangolins are one of the most wondrous living things to walk the Earth. They live in the wild in Africa and Asia, where they frequent tropical forests, dry woodlands and savannah. There are eight species, all mainly nocturnal, of which some spend much of their time up trees and nest in tree hollows, while others prefer to sleep in burrows underground. Most species are about the size of a large pet cat, although some grow to twice this size. Each has a similar shape: a small, pointed head, long snout and a long, strong tail, which can be used to help them climb trees. Their most distinctive feature, however, is that they're the only mammal to be almost completely covered in scales. They're more like a reptile in appearance than a mammal.

The presence of scales has given rise to their alternative English common name: scaly anteaters. Like anteaters, pangolins feast on ants, termites and other small insects, especially insect larvae but, despite their shared diet and moniker, they are not related. Pangolins are closer to the carnivores, like cats, hyenas, dogs and bears, than to other anteaters, although you wouldn't think it by looking at them, especially at their tongues.

Pangolins are toothless so, to help them catch and hold insects, they use their long tongue, which is covered in sticky mucus. It's anchored deep inside the chest cavity and, miraculously, can grow to become longer than the pangolin's entire body. With a single adult consuming an estimated 70 million insects a year, it's clearly a highly effective eating appendage. They've become such ant-eating experts, that they've even evolved the handy ability to close the openings to their ears and nose whilst feeding, so they don't get stung or bitten. They also have sharp, powerful front claws that they use to excavate termite mounds for food.

The word pangolin comes from *peng-guling*, the late-eighteenth-century Malay word for 'roller.' It refers to the way that, when a pangolin is startled or attacked, it covers its head with its front legs and tucks itself into a tight ball, thus presenting a wall of scales to the intruder –

and it works. The scaly armour protects it from the jaws of some serious predators, such as lions, tigers, and leopards. A pangolins will also use the sharp scales on the edge of its tail to lash out, another useful defence. If that's not enough, like a skunk, it can release a noxious-smelling fluid from a gland at the base of its tail.

The main defence for the pangolin, though, is its 'suit of armour' – its scales, made from keratin, the same type of protein that makes up our hair and nails. While hair and nails make up a relatively small percentage of our mass, the pangolin's keratin scales account for about 15 to 20 per cent of the animal's entire weight. They run the full length of the pangolin's body, and their size and shape vary, depending on where on the body they're located. They're arranged in an overlapping pattern, which tapers to fit along the pangolin's long tail. This provides them with protection together with flexibility, and it was this overlapping pattern of scales that caught the attention of a London-based firm of architects.

Our story begins in the 1990s, when the Channel Tunnel, the biggest engineering project ever undertaken by Britain and France, was about to be completed. Constructing a tunnel to connect Britain to mainland Europe had long been a dream for many, including none other than Napoleon Bonaparte, and the first plans to construct such a tunnel were drawn up as far back as 1802. It was, however, nearly two hundred years before the dream finally became a reality. In 1994, engineers completed the 50.45-kilometre tunnel, of which 37.9 kilometres were under the seabed, stretching between Dover in Britain and Calais in France. To this day, it remains the longest underwater section of tunnel in the world.

The Channel Tunnel is built for trains, some carrying cars and lorries, while others are for passengers only, and, it was these new, extremely long passenger trains that needed a specially designed train station for passengers embarking or disembarking in central London. The British government chose the already existing Waterloo Station as the site for this new international terminal and invited architectural firms to bid for the contract. The brief was simple: a streamlined terminal that passengers could pass through with the minimum fuss and at maximum speed. And it wasn't as straightforward a job as many hoped.

Waterloo Station is one of the busiest railway stations in Europe, and it didn't have an awful lot of room to accommodate more trains. All it had was an awkwardly long, thin, and irregular space to the side of the existing concourse, with live electric train lines to the left and the London Underground tunnels beneath. Not only did this space have to include five new train lines and platforms, it also had to house an immigration hall, customs control areas and a dedicated departures lounge – and if that wasn't enough, there was the small issue of the roof.

The roof for the new station would need to take into account the asymmetric site, which narrowed as the new rail tracks left the station. On one side, next to an existing train track, the roof would need to rise more steeply than the other. It also had to cover the entire length of the 400-metre-long platforms to shelter passengers as they waited. Now, if you know London, you know how important shelter is – it rains a lot! And, just to make the whole thing even more difficult, those platforms weren't straight but followed a distinct curve. There was certainly a hefty to-do-list, and, whatever design the architects came up with, it had to look absolutely stunning; so, how did they solve such a mammoth problem? The London-based Grimshaw Architects, working closely with the roofing team of the high-tech engineer Anthony Hunt, came up with a perfect, elegant and beautiful solution, and they based it all on the pangolin.

The architects decided to cover around 50 per cent of the station roof in glass, which would give the roof the wow factor it needed. The curving nature of the site, however, meant that, if they'd built the roof in a traditional way, they would've had no choice but to use hundreds of different-sized glass panels. Many of these, unlike the more commonly used rectangular sheets of glass, would've been highly irregular in shape. With the technology available in the 1990s, this would have been extremely expensive and difficult to construct. The architects needed a solution, and this is when the arrangement of the pangolin's scales came to their attention, in particular, how the scales overlapped, but still allowed for movement. It inspired the architects to adopt what they describe as a 'loose fit' approach to their roof.

Using the cheaper option of standard-sized rectangular glass panels, they designed a system of joints for the roof structure in which the panels fitted together, but could be adjusted, according to the overall shape of the building. This meant that the panels were fitted to overlap at the top and bottom, like the overlapping tiles of a traditional roof. Adjustable brackets and sliding pins, which were also made to a one-size-fits-all specification, held them in place across the entire building. On the sides, the panels were fixed together, using special stretchy seals made from neoprene, the same material from which wet suits are made. These provided the flexibility needed for the panels to follow the distinct curve of the station.

The result was a beautiful and very elegant glass roof that became a stunning symbol for a new era of international train travel. What was even more amazing about

Sunda pangolin
Manis javanica
· Tough scales provide protection
whilst allowing flexible movement.

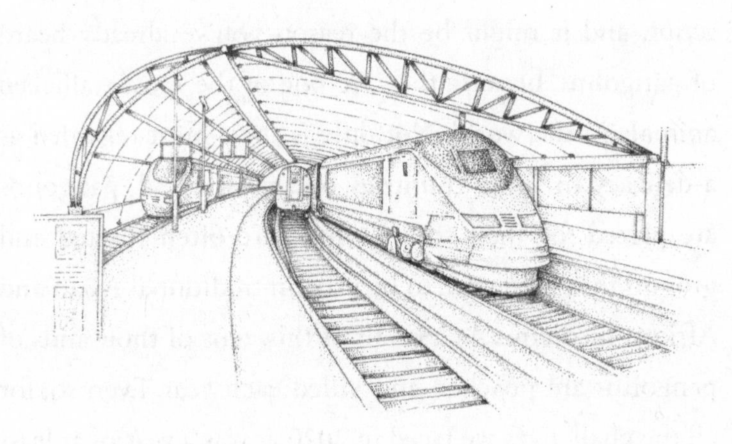
Waterloo International Terminal
· Organic curve of roof
· Built of flexible overlapping glass panes, like pangolin scales

the structure was that it only accounted for 10 per cent of the project's entire cost. For many comparable projects, the roof often comes in much higher than that. The entire station was finished in 1993, a year before the actual tunnel was complete, and it went on to win several prestigious architectural awards. Although, it was decommissioned as the Eurostar station thirteen years later, after the London terminal was moved to St Pancras, its platforms are still used for regional trains. Anyone passing through it would agree that Waterloo International remains a stunning monument to inventive design, and all of this inspired by the scales of the small, insect-eating pangolin – something to think about on your next train journey into London.

There is, however, one sad and rather serious post-script, and it might be the reason you've already heard of pangolins, because they are one of the most trafficked animals in the world. Not only is their meat regarded as a delicacy in some countries but, most of all, pangolins are prized for their scales, which are often roasted and ground into a powder, to be used in traditional Asian and African medicine. As a result of this, tens of thousands of pangolins are poached and killed each year. Even so, for all the challenges we faced in 2020, it was a year of at least one success: the Chinese government officially removed pangolin scales from one of its lists of approved medicine ingredients. The hope is that this type of legislative action will help to protect the pangolin in the future. Although this is a step in the right direction, many pangolin species still remain threatened with extinction. According to the International Union for Conservation of Nature Red List, three are critically endangered, three are endangered and two vulnerable. As a group, the pangolins are right on the edge of a precipice.

28

Swarms of Ants and Mini-Bots

OK, it's question time! What animal has jaws that can snap shut faster than a speeding bullet? If you answered shark or perhaps even crocodile, I wouldn't be surprised. These animals are notorious for their terrifying ability to bite really hard, but the speed with which their jaws clamp shut is nothing compared to our mystery animal. The creatures I have in mind are but a fraction of the size of those massive predators; in fact, they're so small, it's easy to miss them entirely.

They're found in Central and South America and on the Galápagos Islands. They're dark brown, with six legs and big jaws, and come in at just 12 millimetres long. Any idea? Here are a few more clues: we're talking about active hunters, each armed with a venomous sting, and they're one of the fiercest and most aggressive of all the predators in the insect world. Have you got it? I'm talking about ants, in particular, trap-jaw ants. If you've not heard of these little critters before, you're in for a treat, as they've some amazing talents. They're so amazing that robot scientists have been looking to them for inspiration. Let's take a look at their finer details.

During the past 37 million years, trap-jaws have evolved independently up to ten times in the ant family *Formicidae*, but for this story we focus on one species of trap-jaw ants. It doesn't have an English common name, but its scientific name is *Odontomachus bauri*. The group name – trap-jaw ants – comes from having mouthparts that snap shut with extraordinary speed – up to 145 mph – which has earned it a place in the record books as an animal with one of the fastest appendages on the planet. Our ant *Odontomachus*, like all trap-jaw ants, has large and conspicuous jaws that can open wide to nearly 180° and, within 0.01 seconds of being stimulated, can close them 2,300 times faster than the blink of the human eye.

As an aside, it's worth looking briefly at the ant in pole position. The Dracula ant beats *Odontomachus* to the number one spot with jaws that shut at 200 mph in 0.000015 seconds. Of any known animal, it moves its body parts the fastest, but the way it feeds is even more extraordinary. It doesn't feed directly on the prey it catches. Like many insects, including wasps and other species of ants, it catches prey for its larvae, but then obtains essential nutrients by piercing the body of the larvae and sucking out some of their blood, hence the common name. The larvae become food distribution centres for the colony, a form of cannibalism that biologists call a 'social stomach.'

Even though both these species of ants hardly weigh a thing, the speed of their jaws gives them enough force to easily crack through the outer coverings and the armoured exoskeletons of their insect prey. They can decapitate a termite in a split second, well before it has a chance to fight back. The ants can also stun, capture and crack open flies, spiders, beetles, butterflies and other ants. Jaws, however, aren't the only weapons in their arsenal. They also have a powerful sting located at the end of their abdomens, and, since it's not barbed like a honeybee's sting, it can be used over and over again.

If you thought that was enough to be going on with, there's one more surprise in store: the *Odontomachus* ant can use its jaws to launch itself into the air. The bite force is so great that ants have been seen to fly as much as eight centimetres upwards and more than 40 centimetres backwards. When you think about it, for such a tiny insect, that's some impressive propulsion.

Naturalists, who've studied this behaviour in detail, believe that the ants have developed two distinct types of jump. The first's known as an 'escape jump.' The ant places its head and jaws at right angles to the ground, then slams its face straight down, causing the jaws to release with a force about 400 times its bodyweight. This launches it ten body lengths up in the air, and sometimes more. It's a handy trick to escape the probing tongue of a hungry anteater.

The second type of jump has been called the 'bouncer jump' and often occurs when an ants' nest is invaded. Trap-jaw ants live in colonies made up of multiple nests, each containing around 200 ants. Unlike with other ants, these are not in mounds of soil or underground, but are built in flimsy leaf litter, so, they're vulnerable to attack. If an intruder enters, one of the ants will bang its jaws against it, in the hope of bouncing it out. Because of the immense power of the ant's jaws, this action results in the ants themselves being launched horizontally in the opposite direction. This cleverly puts some distance between the ant and the intruder. In fact, when a nest comes under attack, it's not unusual to see a group of ants launching themselves off in different directions at the same time, like popping popcorn. Scientists believe this cooperative colony behaviour is another tactic employed by the ants to confuse potential predators.

How, though, does the spring mechanism work? Imagine there's a worker trap-jaw ant searching for food.

As it moves over and under the leaves and other debris, its giant mandibles – which are the two moving parts of its jaw – are fixed in an open position. The ant can hold them in this way because the muscles in its head contract and pull the mandibles apart, while a pair of latches hold them in place. This turns the jaws into a spring-loaded catch mechanism, full of potential energy ready to be released.

It works like this. Take a bow and arrow, but put the bow to one side. If there is no bow, you only have your arm to launch the arrow through the air. You throw it as hard as you can, but the result is a little disappointing. The arrow doesn't go very far. Now, you've got the bow back in your hands. If you've ever held one, you'll know that it's a bendy and pliable structure. As you stretch back the bowstring, the limbs of the bow become stretched and tense. This means they're storing up a mass of potential energy. When your trigger finger releases the arrow, all this energy sends it speeding towards your target – bullseye!

The ant has a similar mechanism. Once its large muscles have been contracted and the latches have locked its mandibles in the open position, the ant's jaw become like the stretched-out bow, full of energy, waiting to be released. When tiny trigger hairs on the inner edge of the mandibles come into contact with potential prey, they signal the latches to release the jaws, just like an archer

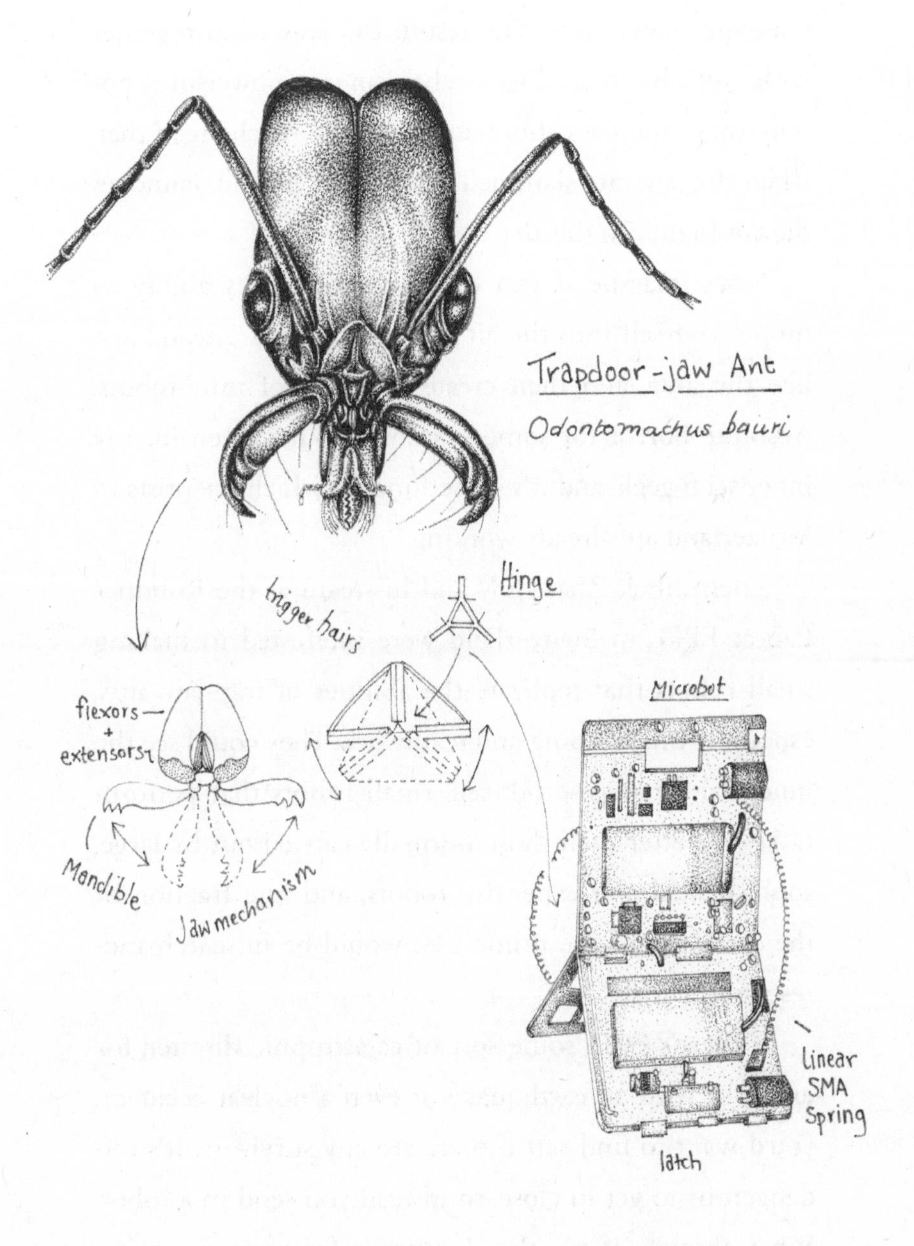
Trapdoor-jaw Ant
Odontomachus bauri
Hinge
trigger hairs
Microbot
flexors
+
extensors
Mandible
Jaw mechanism
linear
SMA
Spring
latch

releasing their arrow. The result: the jaws snap together with explosive force. This mechanism is so powerful, it not only snaps the jaws shut, but is also the mechanism that, when the jaws are slammed against the ground, launches the ant high into the air.

Now, imagine if you could combine this ability to propel yourself into the air with working as a team, just like the ants, and then create a swarm of mini-robots. Absolute horror for some, no doubt, but heaven for my inner sci-fi geek, and it's something on which scientists in Switzerland are already working.

Zhenishbek Zhakypov and his team at the Robotics Lab at EPFL in Switzerland were interested in making small robots that replicate the abilities of trap-jaw ants, especially the jumping and teamwork. They could see the potential for mass-produced, small robots that perform tasks far better than those normally carried out by large, sophisticated and expensive robots, and at a fraction of the cost. One of the prime uses would be in search-and-rescue scenarios.

If there's been some sort of catastrophic disaster, for example, with an earthquake or even a nuclear accident, you'd want to find out if there are any survivors. It's too dangerous to get in close, so instead you send in a robot. What, though, if it's also dangerous for your expensive, high-tech machine? Maybe the terrain is too rough, maybe it's too hot, or maybe your robot is just too big to go looking through the rubble. The answer could be to deploy lots of cheap, easy-to-make, small robots to swarm through the disaster area, each doing a different job, but working as a team. If they're small and lightweight, they could easily reach areas that other robots can't, by slipping through small cracks, and, if one of them became damaged in the process, it wouldn't matter. There would be many others to take its place. So, inspired by the trap-jaw ant, the researchers set about creating such a robot, and they used an ingenious construction method.

You may have seen the *Terminator* films about machines that look like humans, which are sent from the future to assassinate their enemies in the past. The scary thing about some of these humanoids is that they're made out of a material that keeps returning to its original shape. No matter what you do to it – if you shoot or try to blow them up – they just keep coming back. Science fiction? Well, no. It could soon be science fact.

The research team wanted to build its small robots out of a similar flexible, smart material, which would be able to keep its shape. To make this, they layered several different sheets on top of each other, each with a different function. The first layer, known as the 'actuator layer', moved the robot, just like a human muscle moves your arm or leg. The team used a substance, called a 'shape memory alloy', made out of nickel titanium or nitinol. This is a smart metal that can be programmed, using heat or electricity, to produce mechanical movements when the metal returns to its original 'memorised' shape. It meant the robot didn't need a separate, bulky motor. At only 10 grams, it was small, light and powerful.

The team also added a layer of sensors, so the robots could communicate with their teammates and work out if there were obstacles in their surroundings. On top of this was a layer of flexible material for foldable joints, a layer of rigid electronic circuits for its body, and a layer of rechargeable batteries which meant they could be operated remotely from a distance. Now, they had a smart material sandwich that was capable of powering itself without an external source, as well as having the ability to move in and sense its environment. The team printed rectangular shapes out of this sheet and folded them – like you would an origami paper crane – into a triangular 3D object with three legs, about five centimetres tall. They called these origami robots 'tribots'.

So, you might be thinking, where does the trap-jaw ant come in? Well, the researchers added one other function to their robot. At the point where the tribot's joints folded, they added an adjustable spring mechanism, which would store energy, like the trap-jaw ant's open jaws store energy before they snap shut. This meant that the tribot was capable of all sorts of movements. Like the trap-jaw ant, if it came up against an obstacle, the energy stored in its spring mechanism would enable the tribot to jump vertically or horizontally, or even to do a somersault. The tribot also had different ways of covering ground. If the terrain was flat, it could move in a crawl-like, sliding motion but, if it was bumpy, the joints reacted by flipping the robot over from one leg to the other, in a kind of flick-flack type of walking movement. Having this choice was really important because, when you're small, like a tribot, the bumps and obstacles around you are relatively big. And there was more…

The researchers wanted their robot to have another skill copied from the trap-jaw ant: the ability to work as a team. Just as the ants start jumping together to deter a predator, they wanted the tribots to team up to do things normally performed by more sophisticated robots. They did this by pre-programming each robot with a different task. Some would be workers, using their power to, say, move an object, whereas others would be assigned leader roles, using their energy to look out for obstacles. This way, as a team, the tribots became more intelligent and were able to perform more complex tasks

The scientists see many potential uses for their swarm of small robots. One could be in space exploration. Tribots are lightweight and small, so it wouldn't cost much to place them on a rocket, and the fact that they can be programmed to keep their shape means that tribots could even assemble themselves and start moving around – pretty clever, but maybe a bit spooky. For the moment, tribots remain a research project, but who knows – in a matter of years, we might become accustomed to seeing swarms of tiny robots walking, jumping and working together to help us solve problems, and all thanks to the jaws of the trap-jaw ant.

29

Implants and Shocking Tales

There's a very scary predator lurking in the Amazon, so imagine, just for a moment, that you are a small fish swimming along in one of the slow-moving streams that feed into the mighty river. There's almost no visibility. The brown and murky water is clouded with tannins from the breakdown of countless leaves, but it's OK; even though you can't see, you have other senses and the ability to move away from danger using speed. Weaving your way around obstacles, your senses are tingling: there's danger nearby. And then, suddenly, it happens. ZAP! Within the blink of an eye – in fact faster than that, a mere two-thousandth of a second – you have been stunned. Your muscles twitch and contract uncontrollably, and then you're numb, powerless and unable to escape. Before you even have a chance to regain your senses, you're in the dark cavernous mouth of a large fish. You've become the prey of an electric eel.

For me, the electric eel is one of the most intriguing predators to inhabit the great river basins of South America. This is an animal with an incredible superpower – the ability to create electricity within its own body and to deliver the high-voltage jolts that immobilised our small hypothetical fish, but fish are not its only food. Freshwater invertebrates, amphibians, birds and even small mammals are zapped and eaten. To catch and kill them, the eel follows a precise hunting pattern: it generates high-voltage pulses, delivered two milliseconds apart, to detect and locate its prey, and this causes the victim to twitch uncontrollably. It then senses this movement, and delivers the killer blow – a burst of high-voltage pulses at a speed of 400 per second, which stuns or paralyses the prey. Its ability to shock, though, is not just for hunting.

An electric shock is a handy weapon for defence too. A larger predator, such as a jaguar or anaconda, can be immobilised or, at least, deterred, giving the eel the chance to escape. Electric eels have even been known to leap out of the water to deliver warning zaps to land-based predators. In doing this, scientists believe the eels are working in a highly sophisticated way. By leaping into the air, the eels are able to deliver a much stronger electric shock. In the air, they lose less power.

What's even cooler about these 'electro-maniacs' is that electric eels also emit low-voltage electric pulses, which enable them to communicate with other electric eels, and even join together to hunt in a coordinated group. Eels were once thought to be solitary, until Douglas Bastosat Brazil's Instituto Nacional de Pesquisas da Amazônia spotted more than a hundred of them in a remote waterway in the Amazon rainforest. They were hunting tetras, a small and colourful group of freshwater fishes. The writhing mass of eels started to swim together in a large circles and, as the tiny fish were corralled into tight balls, the eels zapped them with a coordinated barrage of electric shocks. The tetras leapt from the water in an attempt to escape, but many were stunned and swallowed. Dr Bastos reckoned the amount of electrical energy generated during the attack could be strong enough to power 100 light bulbs.

Although we're calling these creatures 'eels', they are technically not eels. The electric eel is in the South American knifefish family, whose near relatives include catfish. With its slender, slippery body – grey-brown above and yellow or orange below – and elongated, wavy anal fin that extends to the tip of the long tail, it looks remarkably like an eel. It's also a *big* fish. It can grow up to two-and-a-half metres in length and weigh more than 20 kilograms.

Electric eels live mainly on the muddy bottoms of slow-moving rivers and streams, preferring inconspicuous shaded areas of the forest. These waters, however, are often choked with decomposing vegetation, which strips them of vital oxygen, so the electric eel has developed a clever adaptation. Although it does have gills, it gets about 80 per cent of its oxygen by breathing at the surface, a bit like the arapaima fish in Chapter 19, rising every ten minutes or so to take a breath. Unlike the arapaima, which has developed a modified lung from its swim bladder, the electric eel absorbs oxygen through large blood vessels in its mouth, a form of breathing called 'buccal pumping,' which is basically 'breathing with the cheeks.' Also, it differs from the arapaima in that it's naked: it has no scales.

Although this is an animal with which we're quite familiar, it turns out that we're still learning about these electrifying creatures. As recently as 2019, scientists discovered that there are, in fact, not one but three species of electric eel native to South America. Of the three, *Electrophorus voltai* has earned itself a top spot in the record books. It has the ability to deliver a devastating electric shock of 860 volts. It's the strongest to be delivered by any known animal, and enough to cause an adult human to suffer a full-blown heart attack. People have been known to drown in shallow water after an unfortunate encounter with one of these characters.

Now, though, scientists are putting this deadly power to use in new medical technology. The electric eel is the inspiration for a new type of power source – a soft flesh-like battery – that might be used on our bodies, and even inside them, to power implants. This means they could be used for devices like pacemakers, which stimulate the heart and help it to beat regularly and correctly. So you could say that this powerful zapping predator is being transformed from a heart-stopper to a heart-saver.

It's not the first time the eel has stimulated scientific breakthroughs. The invention of the electric battery, for example, was inspired by the electric eel. In the eighteenth century, electric eels were captured by naturalists and displayed in European theatres. People paid huge amounts to see them and scientists spent their time studying them in order to understand how they could generate electricity. One of those fascinated by the eel was the Italian physicist Alessandro Volta, who, in 1799, based the design of his electric battery – the first in the world – on the eel's anatomy. You may recognise his name. The pressure that pushes a current through a circuit – the volt – was named after him.

So, how does this all work? The key element in the system is a particular type of cell within the electric eel's body, known as an electrocyte. These cells are thin and disc-shaped, and the electric eel has thousands of them inside special organs that can produce both high- and low-voltage charges. As soon as the eel pinpoints its prey, its brain sends a signal through its nervous system to these special electrocytes, which are stacked in long rows, with fluid-filled spaces between them. To help you picture this, imagine a huge stack of circular pancakes – maybe 10, even 20 – each one covered in delicious syrup. Now turn this stack onto its side so that the pancakes are resting on

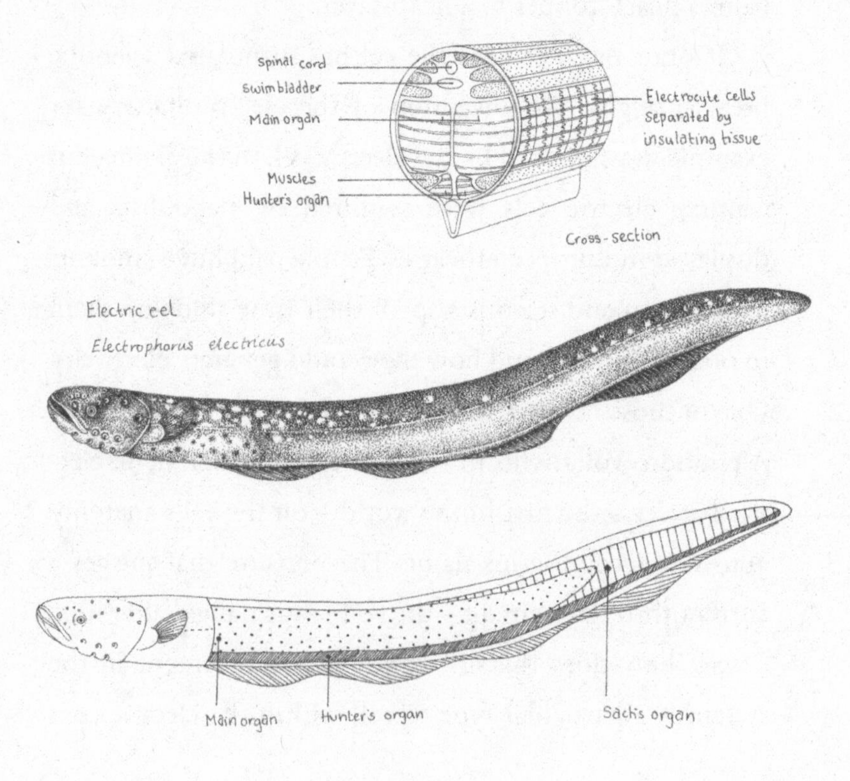

their edges – and this gives you a good idea of what the electrocyte cells look like, along with the fluid in between them. It looks, in fact, a bit like the inside of a car battery.

To understand what occurs within these structures we need to remind ourselves of the physics involved. Everything around us, including our own bodies, is made up of tiny particles. Each particle can have a positive or negative charge. Electricity is the result of this charge moving from one particle to another. When the eel is at rest, each electrocyte pumps out positively charged particles – or ions – through the membrane walls on the front and back sides of the cell. When these positive ions meet in the fluid-filled spaces, they cancel each other out, and nothing happens but, when the eel sends a signal to the electrocytes to act, something amazing takes place. The back of the cell 'flips' direction and lets positively charged particles rush back in. With positive ions now flowing through the cell, from back to front, an electrical current is created.

Inside the eel, there are three special electricity-generating organs producing the high- and low-voltage charges, and they're located in the tail section, which takes up close to four-fifths of the entire body. Inside are thousands of electrocytes, all lined up such that the ions can flow through them. They're arranged in multiple rows, stacked one on top of another, which increases the overall voltage that can be delivered at any one time. Think of how, sometimes, a handheld torch will only turn on when there are lots of batteries lined up inside it – it's the same basic idea.

The first of these organs is called the main organ and it's located on the upper side of the eel, stretching from just behind the head to the middle of the tail. Lying directly beneath this, but running the full length of the tail, is the Hunter's organ. Between them, these two organs produce the high-voltage pulses that the eel uses to stun prey and deter predators. They can deliver hundreds of powerful jolts during a single attack.

The third organ sits behind the other two and is known as Sach's organ. This produces the lower-voltage pulses, which help the eel find its prey and negotiate obstacles in the river. It creates an electric field around the eel's body, which acts like a 'bubble' of electric current. When another animal enters that space, the eel can sense that its electric field has becomes distorted. In this way, it can figure out the other animal's position and even what type of animal it is, and all of this in the dark, murky waters.

To get from organs full of electrocytes to soft, fleshy batteries that can work inside the body, we have to travel to the University of Fribourg, Switzerland, where Michael Mayer, Thomas Schroeder and Anirvan Guha decided to build their own human-made electric organs, stacked full of manufactured electrocytes. They quickly discovered, however, that this was a bigger challenge than they'd first anticipated, mainly because the synthetic electrocytes they created were exceptionally delicate. If one broke, then the entire artificial organ would fail, because the current could no longer pass through it. They decided to simplify the process.

The research team opted for lumps of gel arranged on a sheet. Some of these gels contained water with a high salt content, while others contained water without salt. They were arranged alternately – in other words, one saltwater, one freshwater – in rows with spaces in between. Left in this state the gels did nothing, but, if they were connected in some way, the different concentrations of salt would mean that ions would now be able to flow out of the saltwater gel and into the freshwater one. This, just like the ions flowing in the electric eel's electrocyte cells, would create a small electrical charge… but how to connect the lumps?

The scientists solved this problem by making another sheet on which they placed more lumps of gel, also arranged in rows. If this sheet was pressed face down on the first sheet, the new lumps would fill the spaces, connecting all the lumps together, and allowing the ions to flow. This, our scientists discovered, resulted in an electric charge of up to 110 volts. The drawback, though, was that in order to get this result, they had to use an incredibly large sheet of gel, which was a bit of an issue, because they wanted to create batteries small enough to fit inside our bodies.

The Swiss team then enlisted the help of Max Shtein and Aaron Lamoureux at the University of Michigan, and they came up with an ingenious solution… origami. Yes, you read correctly, it was origami – the Japanese art of folding paper – that held the key. By devising a special folding pattern for the sheet, which allowed the right gels to come into contact with each other and in the right order, the scientists were able to make the battery much smaller without losing any power. This is something small enough to fit into some snazzy body-compatible tech, such as an augmented contact lens – the sort of thing you see in spy movies, where crucial data is beamed directly into the eyes of our heroine or hero. This new battery would be small enough to be used all over the body, for implant devices, such as pacemakers.

At the moment, this fleshy gel-powered battery can only work for around an hour before the levels of ions across all the gels equalise and the battery goes flat. In the future, scientists believe there may be a way of harnessing the chemicals that naturally occur in our bodies to recharge the batteries. If we could do this, we'd be capable of generating our own electricity just like electric eels.

30

Incy Wincy Rescue

I imagine the idea of getting up close and personal with a spider is the last thing many of us would want to do. Arachnophobia – the irrational fear of spiders – is among the most common of phobias, with about five per cent of the world's population suffering from it. I wouldn't say that I have an enormous fear of spiders, but there's certainly something about the way they scuttle about that makes me feel just a little nervous, with the exception, strangely, of tarantulas. Growing up, I was mesmerised by those. I thought they were one of the coolest creatures around, and at ten years old, I really wanted one of my own. That never happened, of course. Can you imagine my mum letting me have a pet tarantula? Not a chance... Fifteen years later, I finally got to hold a tarantula, in all its eight-legged splendour. I was in the Mojave Desert in North America. The animal handler with me carefully placed the spider onto the back of my hand. I couldn't believe how gentle it was, as she explored my hand with slow and considered steps. Now, I know for a fact that, for some people, this is the stuff of nightmares, but I wasn't frightened at all. And it turns out that many aspects of the anatomy, physiology and behaviour of the versatile spider is attracting the attention of research scientists the world over. In fact, the spider just might save your life.

SPIDERS TO THE RESCUE

One aspect of interest is the way spiders move. Unlike insects, which have three main body parts and six legs, spiders have two body parts and eight legs. They walk by alternating pairs of legs. While two pairs are in the air, the other two pairs are on the ground; but it's the way they move those legs, using the hydrostatic pressure of their haemolymph (blood) system, which could have practical value for people.

With no extensor muscles to stretch its legs, the spider extends them by building up the pressure in its body and pumping fluid into them. Control of the hydrostatic pressure is achieved by changing the heart rate – the faster the heartbeat, the greater the pressure. Pressure is relatively low when resting or walking slowly, but running and jumping requires more oomph, up to eight times the resting pressure. This means that, for their body size, spiders can run very fast and jump really well. The Usain Bolt of the spider world is the giant house spider, *Tegenaria duellica*. It can cover roughly half a metre in a second – one reason it's very good at disappearing behind the sofa. One of the most accomplished jumpers is a newly discovered species of aptly named jumping spider from Australia. It can leap almost 50 times its own body length.

If this is a bit daunting, then a giant spider, with legs measuring 20 centimetres long and heading your way, is probably the stuff of nightmares. Then again, imagine the scene of a natural disaster, like an earthquake or tsunami, and you're lying trapped in the rubble, a giant robotic spider could be a lifesaver.

A team at Germany's Fraunhofer Institute for Manufacturing Engineering and Automation has used a spider as the model for a new robot. Being agile and purposeful, the highly mobile robot could be used in environments too hazardous for humans to enter or too difficult for rescuers to access. It could also be equipped to broadcast live images or track down hazards like leaking gas pipes, and send the information back to rescue teams.

With its long limbs, the spider robot is similar to a real spider, but uses air, instead of liquid, to move its limbs.

The eight legs and the body are fitted with pneumatically operated elastic drive bellows to bend and extend its artificial limbs, while hydraulically operated bellows form the joints and keep the limbs mobile. The combination of hinges and bellows means the legs can move forward and turn as needed. Diagonally opposed legs move together. Bending the front pairs of legs pulls the spider robot's body forwards, while stretching the rear legs pushes it. Like a real spider, it keeps four legs on the ground at all times while the other four turn and ready themselves for the next step. This adds to its stability and means it can move over rough terrain without tipping over. It's even possible to make it jump.

The various components required for locomotion – the control unit, the valves and the compressor pump – are stored away inside the main part of the robot's body, which can carry various measuring devices and sensors as and when required. The spider robot is also designed to be very light. The team built it using a 3D printing process called 'selective laser sintering' (SLS), in which thin layers of a fine powder are applied, one at a time, and melted in place with a laser beam. This allows for complex plastic shapes, as well as lightweight components, to be produced.

The team have produced a prototype of the robot but, as they point out, the body could be adjusted for different scenarios, with specialised sensors for detecting various chemical leaks, radiation monitors or even sound sensors and video cameras for search-and-rescue missions. 3D printing also helps to keep the costs down – but that's perhaps not what you want to hear if you're petrified of spiders.

CARTWHEELING SPIDER

Professor Ingo Rechenberg, a German engineer at the Technical University of Berlin and one of the founding fathers of bionics, is studying a somewhat different form of spider locomotion. Although an engineer, he has a passion for deserts and the animals that live in them and, in 2009, he discovered an unusual spider in the heart of the Erg Chebbi Desert in Morocco. It's an oddball among spiders, for it's an acrobat that avoids predators by tumbling to safety; in fact, it's the only known spider to be able to double its speed by actively cartwheeling. It can cartwheel uphill, downhill and across level ground, so it became known generally as the 'cartwheeling spider.' As a new species, it was given the scientific name *Cebrennus rechenbergi*, after the man who discovered it, but Rechenberg prefers to call it by another name: the 'flic-flac spider,' after the German description for an acrobat performing a backward somersault. Desert spiders in Namibia that roll

Flic-flak Spider

Cebrennus rechenbergi

themselves up and use gravity to roll down sand dunes to escape parasitoid wasps have been known for some time, but the flic-flac spider is something else entirely.

It's a medium-sized (two-centimetre-long body) huntsman spider, a family of fast moving, long-legged species that have a superficial resemblance to crabs, hence yet another common name – giant crab spiders. They tend to rundown their prey, and the flic-flac spider hunts moths at night. During the day, it hides in tube-shaped pits lined with silk beneath the sand, but what really captured Rechenberg's attention was its extraordinary movement. As a bionics expert, who applies methods and systems found in nature to the study of engineering systems and modern technology, the spider gave him the idea of creating a cartwheeling robot. It could be used in agriculture, on the deep-sea floor, and even on the planet Mars.

Mars is a terrestrial planet, like the Earth, and many of the rocks and minerals discovered there are also found here. The surface is dry and dusty, rugged with massive craters to the south, but flatter with dry riverbeds and basins in the north, all covered with scattered boulders. Ice caps, that grow and shrink with the seasons, occur at both poles, so exploring the surface of the 'red planet' with robots is challenging.

Working on a solution, Rechenberg set about designing a 25-centimetre-long model of a spider robot that could mimic the flic-flac spider. Whilst he believes this would be ideal for navigating the harsh surface conditions of Mars, it will need to have more stamina than the spider. According to Rechenberg, if the flic-flac cartwheels more than four or five times a day, it can die of exhaustion.

To design his robo-spider, Rechenberg worked with German automation company Feston. The engineers there

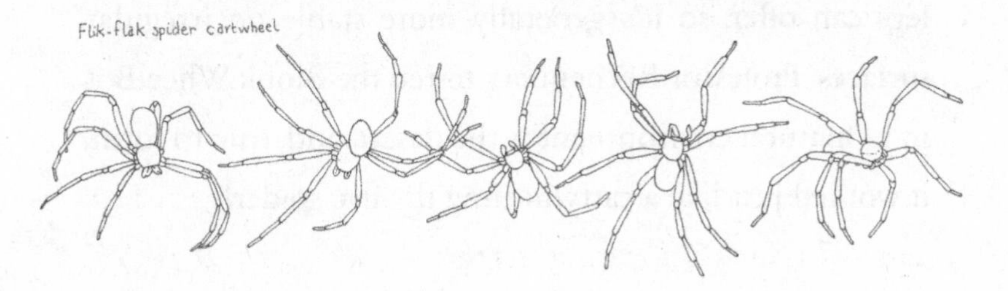

named it the BionicWheelBot, and it looks like something from another planet. It's much bigger than the real thing, of course – about the length of an umbrella – but, like the real spider, it has eight legs, which are controlled by 15 motors within the knee joints and body. It uses six of its legs when walking, but, when it's time to roll, it does a somersault with its whole body, tucking in six of its legs and using the remaining two to push off the ground with every rotation. Thanks to its integrated sensors, the robot knows its exact position, and when to push while rolling. Like the cartwheeling spider, this robot spider is much faster at rolling than walking. It can move itself forward even when on rough terrain, and it's this ability that could make the robot extremely efficient at navigating the tough conditions on Mars.

But why is this somersaulting so important? Well, a robot with a tumbling or rolling motion enables more of its body to be in contact with the surface at any one time. This spreads its weight over a larger area, allowing for greater grip and weight distribution than wheels or legs can offer, so it's generally more stable on irregular surfaces. Professor Rechenberg tested the BionicWheelBot in its 'natural environment' – the desert, and, true to form, it worked just like a cartwheeling flic-flac spider!

ARTIFICIAL SPIDER SILK: JUST LIKE THE REAL THING

One of my earliest memories is of watching a spider weaving its web. I'd seen spiders doing this many times, but this particular spider made an enormous structure, about 60 centimetres across, easily the biggest I'd ever seen. What's more, the insects it had caught were enormous. The web itself looked so delicate and fragile that you'd think they would smash right through, but they didn't and they were trapped.

Spider silk is amazing stuff. Weight for weight, it's about five times as strong as steel, yet is finer than a human hair, and can be stretched several times its length before it breaks, so scientists have had the goal to synthesise it artificially. They know its genetic make-up, and have identified the genes responsible for the production of spidroins, silk proteins that are the building blocks of silk. To make artificial spider silk, however, spidroins must be kept soluble at concentrations as high as 50 per cent, weight for volume, and then quickly converted to threads. People who make proteins industrially spend a lot of time, money and energy preventing proteins from folding up and clumping together, and here again they are turning to spiders.

Spiders are very good at keeping proteins soluble in high concentrations, and scientists in Sweden have been tapping into that with a process that makes kilometre-long fibres of artificial spider silk. Professor Anna Rising at the Swedish University of Agricultural Sciences in Uppsala and Professor Jan Johansson at the Karolinska Institute in Stockholm have carried out the work. It's a complex process, but they found a way to produce new spider silk proteins in such a way that they can be kept in solutions in concentrations as high as 50 per cent, just as they are in the silk-producing glands of spiders, but, importantly, the silk is synthesised by bacteria.

To create these fibres, the researchers mimicked a spider's spinnerets by pumping the protein solution through a capillary (extremely narrow tube) into a low-pH solution. The resulting artificial fibres have the same structure or conformation as natural spider silk, and bear stress and strain in a similar way. The scientists are now exploring ways of turning these fibres into designs and three-dimensional structures for research and clinical applications.

DRUG DELIVERY

In the meantime, at Nottingham University, UK, a spider expert and a chemist have produced a synthetic spider silk that has antibiotic properties. This could be useful in both delivering medical drugs and for closing open wounds with a reduced risk of infection. Spider silk has many advantages as a wound dressing: it's non-toxic, biodegradable, protein-based and, unusually for a protein from a different organism, it is not known to cause any sort of inflammatory or allergic response in mammals. It has been used as a wound dressing for centuries. The ancient Greeks and Romans used to apply it to the wounds of soldiers. A mixture of honey and vinegar was used as an antiseptic to clean the wound, and then wads of spider silk were used to stem the bleeding.

The Nottingham team began with silk they'd cloned using bacteria, which contained special amino acids not normally found in proteins, which can be used to form chemical bonds with drugs or other useful molecules. The silk proteins join together to create fibres up to a metre in length, mimicking what happens as they are extruded from the spider's silk gland, similar to the Swedish experiment. Several different types of molecules are attached to the silk, including the antibiotic levofloxacin, a drug commonly used for treating bacterial infections, and also a molecule which is fluorescent, enabling the degradation of the silk dressing to be monitored when placed on wounds.

To attach the molecules, they are 'clicked' into place inside a solution of synthesised spider silk before the proteins are turned into the actual strands. So what you end up with are silk fibres decorated with slow-release antibiotics. One potential use, the team suggests, could be in the treatment of slow-healing wounds like diabetic foot ulcers. The controlled release of antibiotics means that infection could be prevented. Tissue regeneration and wound healing could also be accelerated by using mesh formed from the silk fibres, which would act like a temporary scaffold to which cells could attach. The silk then naturally biodegrades. It's amazing to think that the spider's web which so fascinated me as a child could provide the inspiration for such revolutionary advancements in human medicine.

SPIDER STREET SIGNS

Has this ever happened to you? You're sitting quietly, and suddenly there's a loud thud at the window. When you turn to look, all you can see is the dusty imprint of a bird. What looks like dust is actually powder, a substance that protects new feathers, and these collisions are surprisingly common. Ornithologists estimate that, in the USA alone, more than 350 million birds die after hitting windows, walls and other human-made structures each year, making this cause of death one of the greatest threats to the lives of birds. Their saviour, at least as far as glass is concerned, could be the spider and its web.

Spider's webs are one of those magical spectacles of Nature. Growing up in England, one of the best things about an early morning in autumn is the mist hanging over the fields, and spider's webs, sparkling with dew, dangling from bushes and trees. Some of these webs, especially those made by large garden spiders – the orb-weavers – can reach a metre in diameter, but this is tiny compared to a spider living in Madagascar – Darwin's Bark Spider – which weaves a web that can measure up to 25 metres across: that's the same length as your local swimming pool. And, the remarkable thing about these webs is that they are generally intact. Birds flit here and there, chasing insects, but they don't blunder into the fine mesh. How come?

It turns out that orb weavers decorate their webs with ultra-violet (UV) reflective threads called stabilimenta. We humans can't see UV light, but birds can, and research has shown that these UV-reflecting threads reduce the number of large birds that crash into their webs. Now a German company has mimicked the UV pattern of an orb-weaver spider's web, using it to coat glass windows and create a surface that deters birds from smashing into it.

The idea began to develop in the late 1990s, when Dr Alfred Meyerhuber, a German attorney and amateur naturalist, read an article in a magazine about orb weaver spiders and their use of stabilimenta. He mentioned it to his good friend Hans-Joachim Arnold, the owner of Arnold Glas, a manufacturer of insulated glass products in Remshalden, Germany. Meyerhuber wondered whether the characteristics of spider's webs could be transferred to glass and, in the process, reduce the number of bird collisions. Arnold was intrigued.

With the popularity of expansive windows and glass walls in modern buildings, and millions of birds being killed worldwide in collisions with window glass each year, there was no time to lose. At first, the Board of Directors took a little convincing, but Arnold managed to persuade them and put his company to work developing a product that would have the same UV-reflecting qualities as spider silk. They worked with their sister company to develop a patterned coating for glass, which is only visible to birds that can detect UV light, and barely perceptible to the human eye.

The companies tested many different coating types and patterns, and they found that a patterned, rather than completely solid, coating made the contrast between the treated and untreated areas more intense. The coated parts reflected UV light while the area sandwiched between two layers of glass absorbed the UV light. The two functions together enhanced the reflective effect, and, although the pattern on a spider's web inspired the initial idea, in order to make the application practical, the team designed their own pattern for the window coating. But was it effective?

Scientists at the Max Planck Institute for Ornithology, Germany, independently tested the glass. A variety of bird species were released inside a nine-metre tunnel, with two glass panes at the far end – one was a control pane made of standard glass and the other was a pane of the UV test glass. More than a thousand test flights were carried out, in which the birds were released and tried to fly out of the tunnel through one of the perceived openings. To protect the birds from actually striking either pane, a net was suspended in front of them, so no birds were harmed during the tests. The researchers recorded the flight path of each bird and the results showed that the birds chose the path towards the standard float glazing significantly more often than the UV test glass. The researchers concluded this was because the UV marker was perceived as an obstacle by the birds. More recently, the company has worked with the American Bird Conservancy and is taking part in their testing programme.

There are other products and other types of glass windows available, which will deter birds, but, in this instance, it's the orb weaver spider's web that has inspired the design. So, next time you run away from a spider in the bath or prepare to flush it down the plug hole, just stop and spare it a thought – do you really want to soak a creature which will help save the lives of billions of birds?

'The machines aren't coming: they're already here'

You know what? Even I have to admit, that's a rather ominous note to conclude with but, in many ways, it perfectly represents the final and ultimate chapter of biomimicry. Before we get into the nuts and bolts of that, though, I'd like to take a moment to extend a heartfelt personal thanks – to you – for taking the time to read this, my first book. I've always found there to be an intangible universality when talking about animals that connects us in a way that transcends language and culture. I don't know what it is exactly, where this feeling germinates from. Perhaps, it's because we so often see ourselves as separate to all the living species with which we share the planet, so when presented with its sheer wonder and beauty we have no choice but to submit to the inalienable truth: we are one with the Earth, and the Earth is one with us.

By now, I can imagine that your mind is buzzing with your own ideas as you look to nature through the lens of bio-inspired ingenuity of the highest calibre. I wonder, what was your favourite story? Perhaps it was the intrepid adventures of space-travelling tardigrades or the shape-shifting, colour-changing antics of octopuses? I know that the alien-like egg-laying of parasitoid wasps was undoubtedly one of the more gruesome tales, but I have a feeling even they got some love. If you want to share your thoughts on any of these chapters with fellow readers like yourself, go ahead and join the conversation using the hashtag #30AnimalsBook.

So where do we go from here? Well, I think there's no doubt that the incredibly inspiring nature of biomimicry will continue to be a core part of how we humans design and create products of the future. One aspect of that future that I find fascinating and near impossible to ignore is the development of artificial intelligence. AI, as we more commonly refer to it, is part of a wider discipline of computer science that today largely encompasses the study of machine learning and artificial general intelligence. Although these areas aren't yet a reflection of true artificial intelligence, they are the fledgling sciences that I believe will undoubtedly lead to a self-aware non-biological intelligence.

Sounds kind of scary, huh? Especially if what comes to mind is a vision of a world overrun by Skynet Terminators and Hal 9000s. A troubling thought. Although I've recently decided that the only way to overcome this fear and uncertainty is for we the public to engage with the development of AI and try our best to have a basic yet informed understanding of how it all works. Considering the topic of this book, an appropriate starting point is to see artificial intelligence as a direct result of biomimicry; after all, it's a branch of science that aims to learn from Mother Nature in order to mimic the intelligence of humans, and other species too.

According to many pundits within the field of computer science, there are said to be three key building blocks to artificial intelligence: input, action, and processing; over the years, we've made huge strides in the development of each. I like to think of 'input' more along the lines of being able to sense the world around you. Animals have always had us beat at this game; whether it's the pin-sharp eyesight of raptors that use their impeccable sight to lock on to prey, or the ability of elephants to detect the deep imperceptible rumbles of infrasound that lead them to the rainwater of distant thunderstorms. Compared to us, many animals come equipped with super senses. In the machine world, this ability to sense also comes in many impressive forms. State-of-the-art cameras easily go beyond the slither of visible light that we see, and can visualise across the entire electromagnetic spectrum to pick out, not only galaxies, but also planets, light years away from Earth. Detecting sound too, for machines, is a mere cake-walk. Even the perception of smell, a sense so complex it has the power to transport our minds to an exact time, place and emotion, is being excelled in the field of bioinformatics. In fact, scientists are already working to develop an electronic nose that replicates the olfactory perception of dogs. Considering the mind-blowing ability of canines today that can be trained to detect the early stages of cancer, as well as positive Coronavirus patient infections from nothing but a sweat sample, this has the potential to revolutionise medical diagnostics and the entire global healthcare system. All thanks, to humankind's oldest and most loyal animal friend.

When it comes to 'action' I put this in the category of movement, and whether it's a cheetah's ability to sprint up to a blistering 80 mph, or how a gibbon can skilfully swing through the upper canopy in death-defying leaps 60 metres above the ground, there are many excellent examples of locomotion in the animal kingdom that would leave us in the dust. Until recently, this was an area that machines weren't necessarily so good at. In fact, robotic attempts at locomotion have over the decades been at the very least somewhat comical, if not laughable; but let's keep that between us, shall we? I'd prefer to keep myself well and truly off the AI naughty list.

In 1973, WABOT-1 became the first robot to successfully walk like a human, although each step took an agonising 45 seconds to complete. Then we have the ASIMO. Created by Honda in the year 2000, not only can this robot walk, but it can also recognise moving objects in its surrounding environment, as well as sounds, faces and gestures which enable it to interact with humans. During an industry demonstration in 2006, however, its attempt to climb a small set of stairs resulted in one of the most iconic 'robot fail' moments to ever grace the internet. In front of a gasping crowd, it toppled over into an expensive malfunctioning heap. The debacle only got better after stunned stage technicians hurried over to place a screen around ASIMO in what some have described as a futile attempt to 'protect its dignity'. Fortunately for us, though, a keen audience member, watching the demo live at the time, uploaded the whole clip online, which has been reposted many times over to amass millions of views. In all fairness, though, we shouldn't be too hard on ASIMO;

to its credit, it was a walking robot that has paved the way for bipedalism in many other humanoid machines.

Since then, things have moved on quite a bit. Seeing as we're on the subject of viral videos, prepare to have your jaw hit the ground. You may have come across this one before, but have a search for the Boston Dynamics creation, 'Spot.' This quadruped robot moves just like a real dog. In fact, it's probably one of the most advanced lifelike machines you'll have ever laid eyes on. It even self-corrects its posture when testers have tried kicking it over, although I'm not sure how I feel about kicking robots. Even that last sentence raises its own questions and a whole new line of ethics. Should we, 'care' about robots? Perhaps it's something we're not all too concerned with right now, but that might change once the final key building block of artificial intelligence crosses the conclusive threshold into self-awareness.

This bring us nicely onto 'processing' or, as I like to call it, brain power. For me, this brain power falls roughly into two categories: the ability to calculate and process information, and the ability to make self-directed decisions. Computers have been masters of the former for quite some time now. Between 1943 and 1945, the wartime Colossus machines – the world's first programmable, electronic, digital computers – proved invaluable in helping British codebreakers at Bletchley Park decipher huge amounts of high-level military intelligence from the German High Command, a processing task that was, and is, far too advanced for our human brains to complete unaided. By 1997, we were using computers to wage a different kind of war, a war of the mind, when for the first time Garry Kasparov, a world chess champion, was beaten by IBM's 'Deep Blue' super-computer. A personal favourite of mine took place more recently in 2016, when, in an emotionally stirring tournament, 'AlphaGo' developed by Google's DeepMind project beat the 18-time world champion Lee Sedol in the 2,500-year-old Chinese board game Go; an abstract strategy game so complex it's said to have more position combinations than there are atoms in the entire universe. All games aside, it's the harnessing of this processing power that's enabled us to discover new medical drugs, has helped us send humans to the moon, and will eventually take us to Mars, and beyond. But for a computer to gain a real sense of cognition in the way we related to it, and become more than a sum of its parts, will require more than simple number crunching.

Intelligence, and I mean true intelligence, is the computing Holy Grail. In the words of Edward Feigenbaum, an eminent interdisciplinary computer scientist of our time, 'The thing that we call AI – computers doing intelligent things – is the manifest destiny of computer science.' I'd even go one step further and add that this is, in fact, the manifest destiny of humanity.

While a lot of us might struggle with the concept of a future lived alongside smart machines, in many ways we're already surrounded by intelligent computers. We now have cars, fridges, TVs and entire homes which have some level of 'intelligent' computing interface built into them. These machines, to which we entrust so many aspects of our lives, can even – thanks to the internet – 'talk' to other networked machines. In these examples, however, computers are simply doing what they've been programmed to do. To go beyond their programming, and perhaps even become self-aware, machines need the capacity both to learn about their world and to go a step further to make an informed decision to perform a task that's contextual to what they predict will happen in the future, based on experiences of the past. This is an important element of intelligence that's seen in humans as well as other living organisms including primates… and birds too. One example I was pleasantly surprised to learn about a few years back while watching the BBC documentary *The Life of Birds*, was how super-smart crows in the Japanese city of Akita had discovered an ingenious method of cracking hard-to-open nuts using the wheels of cars. Perched patiently on the traffic-light signs of pedestrian crossings, these birds remain steadfast, only moving once they see the lights change to signal oncoming cars to stop. As people start crossing, the birds too spring into action, swooping down to the ground to carefully place their hard-shelled nuts on the ground, before retreating once more to the safety of their perch. Even without the cue of pedestrian activity, they can still be seen repeating this action. Once the cars start moving again the weight of the passing vehicles does all the hard work for our crows. This car-using behaviour is thought to have originated at a driving school in Sendai, Japan, in the 1970s, from where the knowledge started spreading to birds in the surrounding areas. It demonstrates wonderfully how animals can, indeed, comprehend a complicated series of events rooted in the past to solve puzzles that require future thought. In fact, other research studies have shown that crows are just as intelligent as many seven-year-olds, if not more so. So much for calling someone a 'birdbrain.'

This ability to learn from previous experiences is a quintessentially biological behaviour. So, can this be replicated digitally, and if so, how? Well, the answer to that question might lie with something called genetic algorithms and interestingly this takes us back to where it all began, with Charles Darwin and his good old-fashioned theory on natural selection.

Genetic algorithms, in case you're wondering, are a type of search heuristic – to you and I this translates more simply as a 'computer based problem-solving shortcut.' These genetic algorithms are directly inspired by Charles Darwin's theory of evolution through natural selection. They're based on standard computing algorithms which themselves are in essence a well-defined set of instructions that help computers more easily solve complex problems. Genetic algorithms however are designed to take advantage of biologically inspired processes such as DNA mutation, chromosomal crossover and natural selection. Now, if all this technical jargon is making you want to stick your overheating noggin in the freezer, don't worry I'm still wrapping my head around the concept too – but you can think about it like this. Much in the same way that polar bears were able to adapt to the challenges of a new and extreme environment – thanks to a mutation in the colour of their fur – the ability for algorithms to mutate and produce 'offspring' with other algorithms allows them to also adapt and go beyond the limitations of their initial programming. I don't know about you, but this sounds a lot like 'thinking' to me.

Artificial intelligence is clearly still in a state of embryonic infancy, but it may well be that genetic algorithms are one of the advances that takes this fledgling technology to the next level. Yet the question remains: are we opening a Pandora's Box, the likes of which we can never close? Maybe we should stop while we're still ahead? Then again, there are many arguments to be made in support of this spirited endeavour. To quote Edward Feigenbaum once more, 'If we don't understand what made ourselves excellent, then we don't know enough, to keep ourselves being excellent over time.'

So, who knows, maybe the world of AI is closer than we think, and one day a self-aware machine will find itself writing a similar book to this one. Only this time they'll be telling tales of the humans that made them smarter.

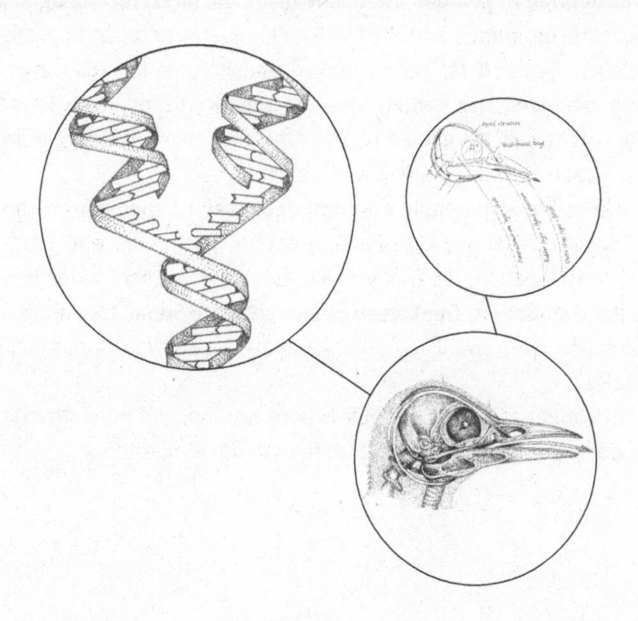

About Island Press

Since 1984, the nonprofit organization Island Press has been stimulating, shaping, and communicating ideas that are essential for solving environmental problems worldwide. With more than 1,000 titles in print and some 30 new releases each year, we are the nation's leading publisher on environmental issues. We identify innovative thinkers and emerging trends in the environmental field. We work with world-renowned experts and authors to develop cross-disciplinary solutions to environmental challenges.

Island Press designs and executes educational campaigns, in conjunction with our authors, to communicate their critical messages in print, in person, and online using the latest technologies, innovative programs, and the media. Our goal is to reach targeted audiences—scientists, policy makers, environmental advocates, urban planners, the media, and concerned citizens—with information that can be used to create the framework for long-term ecological health and human well-being.

Island Press gratefully acknowledges major support from The Bobolink Foundation, Caldera Foundation, The Curtis and Edith Munson Foundation, The Forrest C. and Frances H. Lattner Foundation, The JPB Foundation, The Kresge Foundation, The Summit Charitable Foundation, Inc., and many other generous organizations and individuals.

The opinions expressed in this book are those of the author(s) and do not necessarily reflect the views of our supporters.